BUILDING RESILIENT URBAN COMMUNITIES

COMMUNITY, ENVIRONMENT AND DISASTER RISK MANAGEMENT VOLUME 15

BUILDING RESILIENT URBAN COMMUNITIES

BY

JONAS JOERIN
Center for Development and Environment, University of Berne, Berne, Switzerland

RAJIB SHAW
Graduate School of Global Environmental Studies, Kyoto University, Kyoto, Japan

RAMASAMY R. KRISHNAMURTHY
Department of Applied Geology, University of Madras, Chennai, India

Emerald

United Kingdom – North America – Japan
India – Malaysia – China

Emerald Group Publishing Limited
Howard House, Wagon Lane, Bingley BD16 1WA, UK

First edition 2014

British Library Cataloguing in Publication Data
A catalogue record for this book is available from the British Library

ISBN: 978-1-78350-905-8
ISSN: 2040-7262 (Series)

CONTENTS

LIST OF TABLES

LIST OF FIGURES

ABBREVIATIONS

ACCCRN	Asian Cities Climate Change Resilience Network
ADB	Asian Development Bank
AEE	Assistant Executive Engineer
ALM	Advanced Locality Management
AoRA	Action-oriented Resilience Assessment
APO	Asian Productivity Organization
ASEAN	Association of Southeast Asian Nations
BSUP	Basic Services to Urban Poor
CAP	Climate Action Plan
CBO	Community-Based Organisation
CCA	Climate Change Adaptation
CDCRF	Climate-related Disaster Community Resilience Framework
CDRI	Climate Disaster Resilience Index/Initiative
CMA	Chennai Metropolitan Area
CMDA	Chennai Metropolitan Development Agency
CMWSSB	Chennai Metropolitan Water Supply and Sewerage Board
CoC	Corporation of Chennai
DMA	Disaster Management Act
DRR	Disaster Risk Reduction
EIA	Environmental Impact Assessment
GAR	Global Assessment Report
GFDRR	Global Facility for Disaster Risk Reduction
GHP	Government of Himachal Pradesh
GoI	Government of India
GoIMHUPA	Government of India Ministry of Housing and Urban Poverty Alleviation
GoIMUD	Government of India Ministry of Urban Development
GSDMA	Gujarat State Disaster Management Authority
GSDMP	Gujarat State Disaster Management Policy
HFA	Hyogo Framework for Action
HPEC	High Powered Expert Committee
IDNDR	International Decade for Natural Disaster Reduction
IEDM	International Environment and Disaster Management
IFRCRCS	International Federation of Red Cross and Red Crescent Societies
IMD	Indian Meteorological Department
INR	Indian Rupees
IPCC	Intergovernmental Panel on Climate Change
JICA	Japan International Cooperation Agency
JNNURM	Jawaharal Nehru National Urban Renewal Mission
KU	Kyoto University

LECZ	Low Eleveation Coastal Zone
NDMA	National Disaster Management Authority
NERUDP	North Eastern Region Urban Development Programme
NGOs	Non-governmental Organizations
NIUA	National Institute of Urban Affairs
PAR	Pressure and Release model
RA	Resilience Alliance
SAARC	South Asian Association for Regional Cooperation
UIDSSMT	Urban Infrastructure Development Scheme for Small & Medium Towns
UIG	Urban Infrastructure and Governance
ULB	Urban Local Body
UN	United Nations
UNCRD	United Nations Centre for Regional Development
UNEP	United Nations Environment Programme
UNESCAP	United Nations Economic and Social Communication for Asia and the Pacific
UNHABITAT	United Nations Human Settlements Programme
UNISDR	United Nations International Strategy for Disaster Reduction
UNPD	United Nations Population Division
UNU	United Nations University
USD	United States Dollar
USIOTWSP	United States Indian Ocean Tsunami Warning System Program
WHO	World Health Organization

ABOUT THE AUTHORS

Jonas Joerin
Center for Development and Environment, University of Berne, Berne, Switzerland

Rajib Shaw
Graduate School of Global Environmental Studies, Kyoto University, Kyoto, Japan

Ramasamy R. Krishnamurthy
Department of Applied Geology, University of Madras, Chennai, India

Jonas Joerin is a Research Associate at the Center for Development and Environment at the University of Berne, Switzerland. He worked closely with local communities, governments and international organizations. His research interests are: community resilience, urban disaster risk reduction and climate change adaptation.

Rajib Shaw is the Professor in the Graduate School of Global Environmental studies of Kyoto University, Japan. He worked closely with the local communities, NGOs, government and international organizations, including the United Nations, especially in the Asian countries. He is currently the Chair of the United Nations Asia Regional Task Force for Urban Risk Reduction. His research interests are: community-based disaster risk management, climate change adaptation, urban risk management, and disaster and environmental education.

Ramasamy R. Krishnamurthy is an Associate Professor in the Department of Applied Geology, University of Madras, Chennai, India. During the last two decades he was involved in India's important national programmes on coastal zone management and disaster management. He is currently the Principal of University of Madras Constituent College at Nemmeli Village on the East Coast Road established by the Government of Tamil Nadu in July 2011. As a founder Principal his main tasks are transforming coastal rural students into the mainstream of higher education and equipping the local communities resilient to disasters.

BRIEF INTRODUCTION OF THE SERIES

COMMUNITY, ENVIRONMENT AND DISASTER RISK MANAGEMENT

This series connects academic research to field practice, strengthening the links between the environment, disaster and community. The series will be developed on field evidences and community practices, and thus will provide specific guides to professionals which are grounded in rigorous academic analysis. The series will have specific focus on community-based disaster risk management, urban environmental management, human security, water community, risk communication, climate change adaptation, climate disaster resilience and community-based practices.

BRIEF INTRODUCTION OF THE VOLUME

BUILDING RESILIENT URBAN COMMUNITIES

Recognising ongoing processes of rapid urban transformation in many cities in developing countries, this book throws light on how urban communities in Asian cities, particularly Indian cities, experience impacts of urbanisation and climate change. Related consequences due to unprecedented rural urban migration in many parts of Asia are changing the faces and characters of cities. While such cities try to implement needed measures to safeguard the well-being of their citizens, looming impacts of climate change in form of more frequent and intense natural hazards pose new and additional challenges to their urban communities. In the wake of projected climate change, the research presented in this book connects critical issues related to the general functioning of cities with climate-related disasters and the concept of resilience. The expectation that growing cities in Asia can keep pace in terms of providing the needed basic services to absorb the daily influx of new migrants from outside is all too often impossible. Hence, high numbers of urban poor, fragile infrastructures (sanitation, drainage, road, electricity, etc.) and deteriorating ecosystems are the result of improperly managed cities. We focus in this book on how urban communities of cities in India, particularly Chennai, are capable to absorb a sudden shock in form of a climate-related disaster, such as a flood, storm, drought or heat wave. We further take a pro-active or pro-solution stance that individuals can form collective power which may deliver added-value before, during and after times of disasters. Using the concept of resilience, we apply basic principles of the concept to determine whether an urban community may be affected or damaged during a climate-related disaster or not, and if affected to what extent.

PREFACE

A safe city is a resilient city

Recognising ongoing processes of rapid urban transformation in many cities in developing countries, this book throws light on how urban communities in Asian cities, particularly Indian cities, experience impacts of urbanisation and climate change. Related consequences due to unprecedented rural urban migration in many parts of Asia are changing the faces and characters of cities. Due to increasing demands on infrastructures and urban services, cities in developing countries are often pushed at the edge of collapse already during normal, non-disaster times. While such cities try to implement needed measures to safeguard the well-being of their citizens, looming impacts of climate change in the form of more frequent and intense natural hazards pose new and additional challenges to their urban communities.

In the wake of projected climate change, the research presented in this book connects critical issues related to the general functioning of cities with climate-related disasters and the concept of resilience. The expectation that growing cities in Asia can keep pace in terms of providing the needed basic services to absorb the daily influx of new migrants from outside is all too often impossible. Hence, high numbers of urban poor, fragile infrastructures (sanitation, drainage, road, electricity, etc.) and deteriorating ecosystems are the result of improperly managed cities. Although, it may be important to debate and investigate the causes of such challenged cities, we do not embark on this debate. Instead, we focus in this book on how urban communities of cities in India, particularly Chennai, are capable of absorbing a sudden shock in the form of a climate-related disaster, such as a flood, storm, drought or heat wave. We further take a pro-active or pro-solution stance that individuals can form collective power which may deliver added-value before, during and after times of disasters. Using the concept of resilience, we apply basic principles of the concept to determine whether an urban community may be affected or damaged during a climate-related disaster or not, and if affected to what extent.

To assess the capacity and ability of urban communities to absorb, maintain and bounce-back from climate-related disasters, we developed a tool named Climate Disaster Resilience Index (CDRI) to measure the capacity or resilience of different types of urban communities in India. Key components of the CDRI include: physical, social, economic, institutional and natural dimensions or attributes which are further defined into equal numbers of parameters and variables to calculate resilience levels for selected urban entities.

In the course of this book, individual chapters reveal modified approaches of the CDRI to comfort the scale of different urban communities. While we regard the geographic extent as one key indicator to determine a community, we assessed the resilience of entire selected cities in India, zones (districts within a city) and wards (neighbourhoods) of Chennai. Before assessing the resilience of different urban entities, this book provides an overview about the current state of Asian cities and their exposure to climate disaster risk. This is followed by a review of the concept of resilience applied in urban areas. Since the concept of resilience applied on urban entities includes aspects of urban planning and sustainable development, a brief overview about Disaster Risk Reduction (DRR) implementation in India is provided. In the remaining chapters, the linkage between resilience and DRR is discussed and solutions are proposed how to enhance or build resilience at different urban scales to climate-related disasters.

We recommend this book for everyone as it is written in an easy understandable manner. However, it may be of particular interest to the following: students, scientists, managers in the field of disaster management, urban planners and NGO representatives.

- Undergraduate and postgraduate students, as well as scientists, from natural and social sciences should benefit from new approaches about how to use and practically apply the concept of resilience in the context of cities located in developing countries.
- Managers and officials and planners working for local and regional authorities find practical approaches about how to measure the resilience of their cities, districts, wards, etc. to disasters. Furthermore, solutions are provided in this book about how to enhance the resilience of an urban entity in the context of DRR implementation.
- Representatives working for NGOs can learn about the current approaches to assess resilience in urban areas and among communities. The book also serves as a guidance to know how to enhance the climate-related disaster resilience of urban communities.

We would be delighted if this book finds the approval of the readers.

We are deeply grateful to a wide range of people for their support and assistance in developing this book. We also appreciate great help received from many organisations and institutions. In particular, we are indebted to the help, assistance and support received from the following:

- The Corporation of Chennai and in particular Mr. Natarajan Mathavan, who provided crucial technical and administrative support to facilitate the collection of data, conduction of workshops and organisation of meetings with key stakeholders in Chennai.
- The Japanese Society for the Promotion of Science (JSPS) and Kyoto University with its Global Centre of Excellence (GCOE) and Human Security Engineering (HSE) programmes which provided substantial financial assistance to this research allowing the authors to carry-out various field visits, conduction of workshops and data collection.
- The National Institute of Disaster Management of India and Seeds India which generated technical and logistical support to facilitate the implementation of a CDRI assessment among Indian cities.
- Students and professors (particularly Prof. K.K. Sharma) from the Department of Applied Geology of the University of Madras, India, for giving guidance, support for data collection and critical remarks on the overall approach and conduct of the research.

In addition, deep appreciation goes to all people and organisations who are not explicitly mentioned, but have contributed in many ways to this book.

Jonas Joerin
Rajib Shaw
Ramasamy R. Krishnamurthy

CHAPTER 1

INTRODUCTION

The earth is witnessing dramatic changes due to a growing population which demands more and more natural resources. In particular, fossil fuels have been used as natural resources to provide the energy to run the earth's economy. The consequence of this process is a man-made global warming which has led the planet's temperature to rise over the last 160 years at an unprecedented rate compared to the earth's age (Intergovernmental Panel on Climate Change (IPCC), 2007). The impacts of climate change are likely to aggravate further the rising trends of hydro-meteorological-related disasters, such as floods, storms, droughts, etc. (IPCC, 2007). In combination with growing urban populations, particularly in developing countries, the number of people becoming affected to natural hazards is expected to increase if no action is taken.

CLIMATE DISASTER RISK

To begin, the term 'climate disaster risk' is circumscribing the challenges associated to increasing or changing vulnerabilities of people and objects to climate-related natural hazards. Thus, evolving on the commonly known term 'natural hazard risk', climate disaster risk refers to risks that have direct linkage to disasters. For example, an urban poor community which is located in the dry parts of a river bed is not just exposed to a potential natural hazard risk from the river, but has inherent weaknesses in form of socio-economic disadvantages that create additional risk. In other words, the amalgamation of various types of risk existing in a built environment, which is inhabited by humans, is suggested to be better represented by the term climate disaster risk. One may now question the adequateness of the word 'disaster' in this previous term. This is simply due to the expected loss that could occur in form of damaged infrastructure or loss of human lives which are the key characteristics determining a disaster.

Going back to the various types of risk, urban areas are by nature hoarding larger numbers of risk compared to rural areas simply due to their characteristics as having higher densities of living human populations, use of land and extraction of natural resources (Wisner, Blaikie, Cannon, & Davis, 2004). As a result, urban areas are more susceptible to impacts of natural hazards causing disasters compared to rural areas. Adding to the fact that patterns of natural hazard risk are likely to change due to climate change make the term 'climate disaster risk' a favourable one to describe the before mentioned linkages of various types of risk.

In the context of urban areas, climate disaster risk is increasingly becoming important to be considered in land-use planning and measures protecting the safety of communities, particularly in cities in developing countries. This is due to the following types or drivers of risk that may add to the potential occurrence of a disaster in such urban areas: impacts of urbanisation and changing land-use patterns. Although, impacts of urbanisation may ultimately cause changes of land-use, these two aspects shall be treated separately due to the following reasons: firstly, changes of land-use can also occur among a stable urban population who demands new developments that support changes of the urban landscape; secondly, impacts of urbanisation inflict more than just new or higher demands of land available for development. Especially, in developing countries, urbanisation is often a key factor causing unplanned development in areas that are not yet designated for development. This unplanned development may also often be associated to encroachment of water bodies, such as lakes, protected marsh land, swamps or areas with high ecologic value which are not designated for development or not yet defined in land-use plans. Urbanisation may further cause pressures on the local authorities to provide sufficient civic services demanded by an increasing population. And lastly, urbanisation has direct impacts on the use and quality of natural resources of a city. As in this book, the primary focus is on urban areas in India and, in particular, on Chennai, the various types of climate disaster risk are from now on referred as 'urban risk'.

Urban Risk

According to UN predictions, the urban population has become bigger than in rural areas since around the year 2009 (United Nations Population

Division, 2010). As a result of this urbanisation process, in many parts of the world, especially in Asia and Africa, cities have become vulnerable to urban risk. This urban risk is characterised by issues related to urban poverty, lacking basic urban services, loss of urban greenspace, disrupted ecosystems, exhausted institutions and unplanned development which is further discussed in Chapter 2 in the context of Asia.

Climate Change and Disasters

According to the Intergovernmental Panel on Climate Change, the global warming over the past decades is man-made and has led the global temperature to increase by 0.8°C in the 20th century and is further expected to rise depending on the scenarios and location on the earth by up to 6.4°C in the 21st century (IPCC, 2007). Furthermore, it is expected that climate change or climate-related hazards are likely to increase in the future. Looking at Fig. 1, this scenario is already proving that climate-related disasters are increasing compared to geo-physical disasters which remain stable over the past 60 years.

However, whether the increase of climate-related disasters is attributable to climate change is rather unlikely. Instead rising numbers of floods, storms, etc. are largely due to a rising population that is exposed to these hazards. Since urban areas are by nature densely populated they are

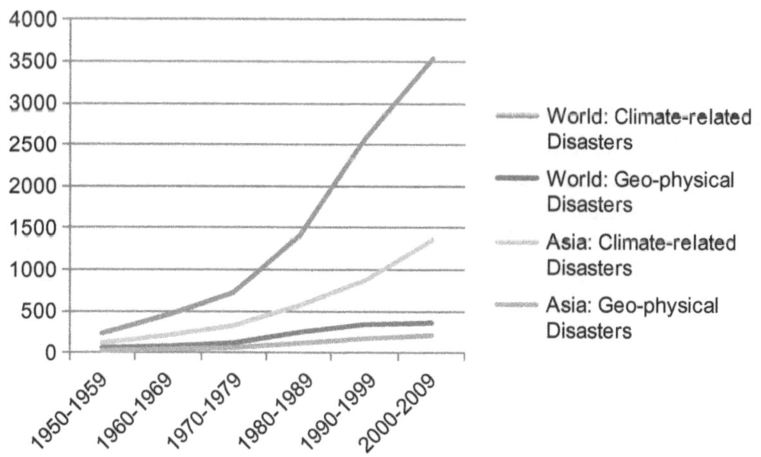

Fig. 1. Increasing Number of Disasters. *Source*: EM-DAT, 2011.

more and more becoming hot-spots for intensive disaster risk compared to rural areas (United Nations International Strategy for Disaster Reduction, 2009), as mentioned above. Hence, the combination of urbanisation, particularly in hazard-prone areas, may increase the occurrence of future disasters in all parts of the world. However, since most of the future population growth will take place in Asian urban areas (United Nations Human Settlements Programme (UNHABITAT), 2008, 2010; United Nations Population Division, 2010), research is needed to understand the processes within cities and their communities on whether they are prepared for future increase of disasters.

CLIMATE DISASTER RESILIENT URBAN COMMUNITIES

Owing to the change of numerous urban patterns to reduce climate disaster risk in urban areas, concepts of sustainable development are needed. Such concepts ought to be integrated and deemed to require holistic development which is reflecting the physical, social, economic and environmental needs of urban communities. The reflection of such elements is crucial in the process of creating safer or disaster resilient communities.

Resilience, Community, Climate Disaster Risk

In a primary step, the associated linkage between climate disaster risk, community and resilience needs to be emphasised. In this book, an attempt is made to link elements of climate disaster risk to the concept of resilience. Thus, it is advocated that reducing climate disaster risk in an urban area requires strengthening the resilience of a community and vice versa.

In recent years, the concept of resilience has become increasingly popular to describe the ability of people, entire systems (e.g. cities), materials, etc. to manage disruptions. Hence, in this book it ought to serve as a theoretical concept that is transformed into practical tools fostering the enhancement of capacities of urban communities. Despite the theoretical character of the concept of resilience, the following chapters aim to demonstrate its ability to serve as an implementation-oriented framework that can address the challenges faced in urban areas in relation to climate-related disasters.

Accordingly, the following key questions emerge which are attempted to be answered in this book:

- How to measure climate-related disaster risk in a spatially defined urban area?
- How to support community actions leading to climate-related disaster resilience?
- How to involve different stakeholders into the process of enhancing the resilience of urban communities to climate-related disasters?

Assessing Community Resilience

Based on studies undertaken by various scholars (e.g. Bruneau et al., 2003; Cutter et al., 2008; Twigg, 2007), concepts and theoretical approaches have been undertaken to define aspects of community resilience. However, detailed and quantitative assessments measuring the resilience of communities have been urged by these authors to be lacking. Hence, in this book, resilience assessments have been developed aiming to address issues of climate disaster risk prevalent in urban areas.

The following resilience assessments are discussed in the following chapters:

- Climate Disaster Resilience Index/Initiative (CDRI) assesses the ability of different types of urban areas (cities and zones/districts/neighbourhoods) to cope and respond to climate-related disasters.
- Action-oriented Resilience Assessment (AoRA) investigates about the perceptions of community leaders to enhance the resilience of an urban area through specific actions related to Disaster Risk Reduction (DRR).
- Climate-related Community Disaster Resilience Framework (CDCRF) assesses the resilience of households to climate-related disaster resilience. Based on physical, social and economic aspects, this framework aims to understand whether households learn from climate-related disaster and take action to increase their resilience or not.

Mainstreaming Disaster Risk Reduction at the Local Level

Aiming to plan for safer communities which are characterised by minimal exposure to disaster risk and high resilience ultimately requires a multi-stakeholder dialogue and, most importantly, cooperation between actors engaged at different spatial scales. Thus, measures supporting DRR need to be effectively integrated into urban land-use plans, policies, laws and regulation to recognise a plethora of risk elements challenging the safety of urban communities. As a result, expected losses of lives and infrastructures due to disasters may be lessened or even avoided.

CASE STUDY LOCATIONS

In this book, various study locations are described which have been selected for resilience studies. In a primary assessment of a city's resilience, a total 13 cities (Aizawl, Amritsar, Bhubaneshwar, Chennai, Delhi, Guwahati, Jaipur, Kanpur, Kolkata, Nagpur, Port Blair, Shimla and Varanasi) with different topography and size were investigated to get an understanding on the overall resilience of Indian cities to climate-related disasters.

In further detailed assessments in Chennai, resilience assessments have been applied at different urban scales. Chennai is located in the state of Tamil Nadu and lies neatly along the Bay of Bengal. The size of the city before its expansion was 176 km^2 which has been increased in October 2011 to 426 km^2. The geographical coordinates of the city are 13°5′2″N80° 16′12″E which indicates a tropical location with a hot climate. Indeed, the average temperature ranges between 24°C in January (coldest month) and 33°C during the hottest months in April and May. The yearly average rainfall is around 1,230 mm with most of the rainfall during the post-monsoon period from October to December. During this time, the occurrence of cyclones is likely possible which have been established over the Bay of Bengal. Therefore, most recorded disasters take place between October to December due to either rainfall-induced floods or devastating cyclones (Revi, 2008). Based on climate change predictions, the city is likely to face serious challenges due to rising sea level (Aggarwal & Lal, 2001) that could be as high as 49 cm during the 21st century. This will not only cause coastal erosion but may also have wider impacts in form of storm surges due to the Chennai's low altitude with an average of 1.5 m above sea level. Furthermore, sea-level rise may cause salt water intrusion into the groundwater that would hamper the supply of drinking water. The low average altitude is due to the city being developed on a former lagoon area and thus no hills exist.

The city has experienced rapid population growth during the past century until today. After Chennai's foundation in 1639 (Muthiah, 2008), the city consisted of scattered villages until the city became a primary port in India during the middle of the 19th century. Following the economic growth, the population rose and trespassed the first million in 1943. After the independence of India from the British Empire, the city witnessed its most dramatic absolute population growth during the 1950s to 1970s (Chennai Metropolitan Development Authority (CMDA), 2008). In the past decade, the city grew only marginally and has now become more or less stable. Chennai's current population stands at 4.6 million in 2011 compared to 4.34 million in 2001 (Census of India, 2011). The consequences of this rapid urban expansion led to the uprising of informal settlements and urban poor areas in particular along the canals and other water bodies of the city. The percentage of the urban slum population was predicted to be around 18.2 per cent in 2001 (CMDA, 2008), however, recent figures indicated more than 30 per cent. Unlike the general decreasing trend of urban poverty (UNHABITAT, 2010), Chennai is still facing increasing numbers. As later discussed in more detail, the consequences of urbanisation also challenge the various basic urban services, such as the provision of electricity, water, sanitation, sewerage and solid waste management.

ABOUT THE BOOK

This book is divided into eight chapters.

Chapter 1 introduces the context of the book emphasising on the emerging pressures due to urbanisation and climate change in cities, particularly in developing countries. To assess the issues associated to rising climate disaster risk in urban areas, the concept of resilience is adopted to investigate the abilities and capacities of urban communities.

Chapter 2 highlights climate disaster risk reflected based on experiences and current situations with focus on the urban context of Asia.

Chapter 3 focuses on the concept of resilience which is applied to understand the ability and capacity of cities and communities in urban areas to manage and respond to climate-related disasters.

Chapter 4 reviews the evolution of DRR as a concept that can support the objective of enhancing resilience in urban areas. Thereby, the focus is on available tools that are providing the legal and financial capacities of Indian cities allowing DRR to be fostered in urban areas.

Chapter 5 presents the development of a CDRI to assess the climate-related disaster resilience of different types of urban areas. Accordingly, a selected number of cities across India and a more detailed analysis of neighbourhoods (zones) within a single city (Chennai) aim to provide an understanding of resilience in urban areas.

Chapter 6 presents an AoRA to assess the perceptions of local decision-makers or community leaders about how to enhance the resilience of urban communities in Chennai and which stakeholders (local government, communities, academia, private organisations and non-governmental organizations) should take the lead in the implementation of resilience-enhancing actions.

Chapter 7 assesses the perceptions of residents through a CDCRF. Based on a household survey, two communities have been examined to understand their current physical, social and economic resilience. In addition, disaster-affected households were asked whether they took action to enhance their resilience.

Chapter 8 discusses the findings from the studies presented in Chapters 5–7 and put into the context of the presented literature in Chapters 2–4 to develop solutions and recommendations for building resilience to climate-related disasters in Chennai.

REFERENCES

Aggarwal, D., & Lal, M. (2001). *Vulnerability of Indian coastline to sea-level rise.* New Delhi: Centre for Atmospheric Sciences, Indian Institute of Technology.

Bruneau, M., Chang, S. E., Eguchi, R. T., Lee, G. C., O'Rourke, T. D., Reinhorn, A. M., Von Winterfeldt, D. (2003). A framework to quantitatively and enhance the seismic resilience of communities. *Earthquake Spectra, 19*(4), 733–752.

Census of India. (2011). *Provisional population totals paper 1: Tamil Nadu. Government of India.* Retrieved from http://censusindia.gov.in/2011-prov-results/data_files/tamilnadu/3.Tamil%20Nadu_PPT_2011-BOOK%20FINAL.pdf. Accessed on October 30, 2011.

Chennai Metropolitan Development Authority (CMDA). (2008). *Second master plan for Chennai metropolitan area, 2026.* Chennai: CMDA.

Cutter, S. L., Barnes, L., Berry, M., Burton, C., Evans, E., Tate, E., & Webb, J. (2008). A place based model for understanding community resilience to natural disasters. *Global Environmental Change, 18*, 598–606.

EM-DAT. (2011). Database. Retrieved from http://www.emdat.be/. Accessed on October 29.

Intergovernmental Panel on Climate Change (IPCC). (2007). Summary for policymakers. In S. Solomon, D. Qin, M. Manning, Z. Chen, M. Marquis, K. B. Averyt, ... H. L. Miller (Eds.), *Climate change 2007: The physical science basis.* Contribution of Working Group I to the Fourth Assessment Report of the Intergovernmental Panel on Climate Change. Cambridge, MA: Cambridge University Press.

Muthiah, S. (2008). *Madras rediscovered.* Chennai: Westland Limited.

Revi, A. (2008). Climate change risk: An adaptation and mitigation agenda for Indian cities. *Environment and Urbanization, 20*(1), 207–229.

Twigg, J. (2007). *Characteristics of a disaster-resilient community: A guidance note*. Disaster Risk Reduction Interagency Coordination Group. DFID: Benfield.

United Nations Human Settlements Programme (UNHABITAT). (2008). *State of the world's cities 2008/2009*. London: Earthscan.

United Nations Human Settlements Programme (UNHABITAT). (2010). *The state of Asian cities 2010/11*. London: Earthscan.

United Nations International Strategy for Disaster Reduction (UNISDR). (2009). *Global assessment report on disaster risk reduction*. Geneva: United Nations.

United Nations Population Division (UNPD). (2010). *World urbanization prospects: The 2009 revision*. New York, NY: United Nations Population Division, Department of Economic and Social Affairs.

Wisner, B., Blaikie, P., Cannon, T., & Davis, I. (2004). *At risk: Natural hazards. People's vulnerability and disasters*. Cambridge, MA: MIT Press.

CHAPTER 2

CLIMATE DISASTER RISK IN URBAN AREAS IN ASIA

Over the past decades, many cities in low- and middle-income countries have experienced unprecedented population growth (Satterthwaite, Huq, Pelling, Reid, & Romero-Lankao, 2007). In particular, cities located in Asia, Africa and Latin America are facing rapid inflow of new dwellers (United Nations Population Division (UNPD), 2010) not only causing spatial physical expansions of urban areas but also serving as economic drivers of the regions beyond the respective urban areas (United Nations Human Settlements Programme (UNHABITAT), 2008). The combination of negative impacts of urbanisation trends, such as more densely populated living environments for people and higher demand of basic urban services, etc. (UNHABITAT, 2008), with projections from the Intergovernmental Panel on Climate Change (IPCC) that the climate is likely to change in the future (IPCC, 2007) brings challenges to the development and planning of cities. As a result, cities with limited capacity to respond to these recent phenomena are expected to suffer from climate-related disasters.

In this chapter, firstly, emerging urban risk is reviewed in relation to pressures from climate change, focusing on urban areas located in the Asian region, particularly India. Secondly, the theoretical concept of resilience is linked to how different types of urban areas are likely to respond and manage climate-related disasters.

IMPACTS OF URBAN TRANSFORMATION PROCESSES AND CLIMATE CHANGE

Urban Population Trends

Around the year 2009, the global urban population trespassed the 50 per cent mark (Fig. 1) where more people live in urban than rural areas

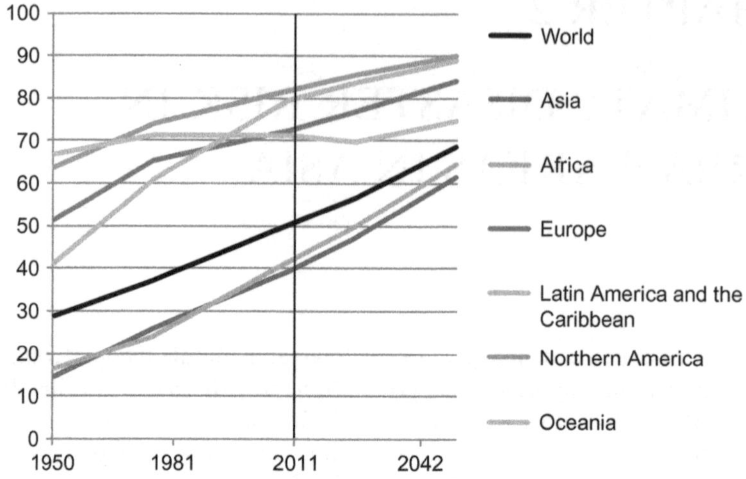

Fig. 1. Global Urbanisation Rates (in per cent), 1950–2050. *Source*: UNPD (2010).

(United Nations Population Division (UNPD), 2010). Accordingly, the projections do not show a change of this growing urban population trend in contrast; it is further expected that particularly cities in countries in Asia, Africa and Latin America will continue to grow in the coming decades and contribute to the global urban population (UNHABITAT, 2008; UNPD, 2010).

Despite estimations that the number of megacities (population higher than 10 million) is going to increase from 21 (12 in Asia) in 2009 to 29 in 2025, their total share of the global urban population will remain at 10.3 per cent in the year 2025 compared to 9.4 per cent in 2009 (UNPD, 2010). Thus, the largest 'receivers' of new residents are small- and medium-sized cities with population sizes below 1 million.

Although, more than 50 per cent of the global population now lives in urban areas, the urbanisation rates of Asia and Africa are still below this mark with 42.2 per cent in 2010, respectively, 40 per cent for the latter (UNHABITAT, 2010). For example, it is not expected that before the year 2026 the Asian region will reach the 50 per cent mark. Although being the largest country on the Asian continent, China has yet to reach this threshold line which is expected to be the case before the year 2014.

The transition from rural to urban living habits in China stands in stark contrast to the second largest nation (by population) in Asia, India, which stands at only 30.2 per cent of people residing in urban areas in 2011 (Census of India, 2011), and is not expected to reach the tipping point before 2044 (UNHABITAT, 2010).

However, India's urban areas are recently experiencing rapid population growth which is highlighted by a high decadal growth rate of 31.8 per cent in the period from 2001 to 2011 compared to only 12.2 per cent increase in rural areas (Census of India, 2011). In comparison to the world average yearly urban growth rate of 1.76 per cent (UNPD, 2010), the urban growth rate in India between 2001 and 2011 was 2.8 per cent (Census of India, 2011). Thus, significantly higher than the measured population growth in rural areas of India that stood at a yearly rate of 1.2 per cent and 1.6 per cent for the total population growth of the nation during the same decade.

Although the total average population growth in India was 17.6 per cent during the period from 2001 to 2011 (Census of India, 2011), urban areas absorbed in average a larger amount of people compared to rural areas. This evidence highlights that urban areas in India are gaining importance, particularly as drivers of economic growth in the respective regions (UNHABITAT, 2010).

While urban areas in developing nations of Asia and Africa have similar patterns of attracting people from rural areas into cities, the rising urbanisation rate in Europe (Fig. 1) is largely due to urban areas merging together and building urban agglomerations or corridors (UNHABITAT, 2008). For example, in the Rhine-Ruhr metropolitan region in Germany, where a large number of former industrial cities merged to an urban area stretching over 4,435 km^2 and carries around 5.2 million inhabitants. Other examples of connecting urban areas in develop countries are between New York and Washington D.C. in the United States or between Tokyo and Osaka (UNHABITAT, 2008). These areas ultimately have a driving role for the economic growth in their regions.

Emerging Urban Risk

Although the population growth of urban areas is contributing and accelerating the economic growth and vice versa – per se positive – the inflow of large numbers of people into relatively small areas (city) within a short period is seen as a serious challenge for urban systems, particularly in developing countries (United Nations International Strategy for Disaster Risk Reduction (UNISDR), 2009; UNHABITAT, 2008, 2010). Despite the beneficial role of cities as strong economic drivers in the region, there are often problems associated with unprecedented urbanisation rates in Asian cities (UNHABITAT, 2010).

Urban Poverty

Although extreme overall poverty in the Asia-Pacific region reduced from 49 to 25 per cent from 1990 to 2005, urban areas particularly in the South Asian region show growing trends of poverty (UNHABITAT, 2010). Accordingly, it was estimated in 2010 that the Asia-Pacific region would be the 'host to over half of the world's slum population' (UNHABITAT, 2010, p. 14).

Fig. 2 shows the percentage-wise reduction of urban slum population in the three most affected continents in the world. Thus, as of today, Africa still has the highest percentage of urban slum population followed by Asia and the Pacific and Latin America and Caribbean. However, UNHABITAT (2008) points out that slum populations are not reducing equally fast in different regions within Asia and the Pacific. Although, the slum population in the South Asian Association for Regional Cooperation (SAARC) to which eight countries (Afghanistan, Bangladesh, Bhutan, India, Maldives, Nepal, Pakistan and Sri Lanka) belong, also commonly referred as South Asia, decreased from around 41.5 per cent in 2000 to 32.1 per cent in 2007 (United Nations Economic and Social Communication for Asia and the Pacific (UNESCAP), 2011). However, as Table 1 highlights

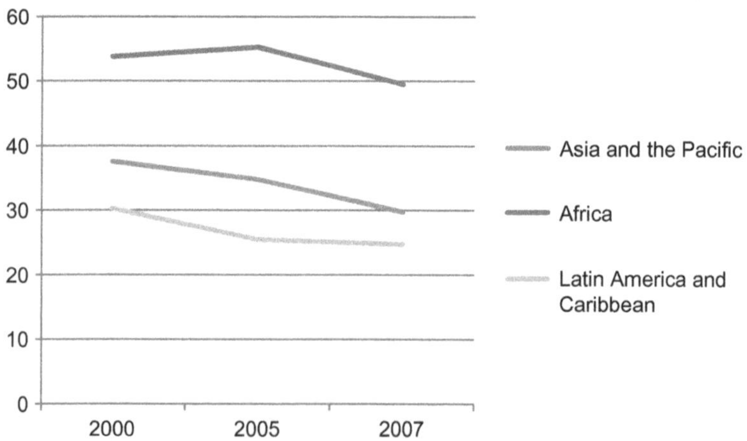

Fig. 2. Slum Population in Africa, Asia and the Pacific and Latin America and Caribbean. *Source*: UNESCAP (2011).

Table 1. Estimated Slum Population in Asia and the Pacific in 2001.

Subregion	Total Population	Urban Population		Estimated Slum Population	
	Millions	Millions	Percentage of total	Millions	Percentage of urban
East	1364	533	39.1	193.8	36.4
South-Central	1507	452	30.0	262.4	58.8
South-East	530	203	30.3	56.8	28
Oceania	8	2	26.7	0.5	24.1
Total Asia	3409	1191	34.9	513.5	43.1

Source: Adapted from UNHABITAT (2003).

the percentages for the South-Central area to which many of the SAARC nations belong was estimated much higher in 2001 with 58.8 per cent (UNHABITAT, 2003).

In absolute numbers, around 513 million urban poor lived in settlements in 2001 (see Table 1) who were lacking basic urban services and sufficient livelihood securities (UNHABITAT, 2003). These figures in Table 1 and Fig. 2 may not fully match each other which points out the difficulties in getting accurate data (UNHABITAT, 2003). Despite the challenges in the determination of percentages of slum populations, the absolute number of slum population is increasing in many countries and has not yet come to a halt (UNHABITAT, 2008).

In the case of India, the number of people living below the poverty line in urban areas has declined from 49 per cent in 1973/1974 to 25.7 per cent in 2004/2005; although, the poverty line defined in India is far below the international standard of USD 1.25 per day. Based on 2004/2005 estimations, the India's average urban poverty line stood at just USD 0.48 per day (INR 19.2 per day) (Government of India Planning Commission, 2009). However, the Planning Commission of the Government of India has proposed to increase the poverty line to INR 965 per month or around INR 32 per day (IBN-Live, 2009) which would equal to around USD 0.65 per day. Not surprisingly, more than 80 million people live in slums in India (Government of India Ministry of Urban Development (GoIMUD), 2011). Accordingly, the average rate of slum settlers in urban areas rose from 21.3 per cent in 1991 to 23.5 per cent in 2001. An example with high above average slum population is Mumbai with up to 54 per cent (GoIMUD, 2011).

The root causes of urban poverty in developing countries are diverse and different depending on the contexts; however, it is argued that the repeated failing of structural adjustments policies in the past have triggered uneven development processes causing urban poor settlements to mushroom (Silva, 1996; Wratten, 1995) particularly in hazard-prone areas of cities (Pelling, 2003; Satterthwaite et al., 2007).

To summarise, the perceived continuing trend of growing urban poverty in the Asia-Pacific region points out that unprecedented population growth rates in urban areas are going beyond the capacity of urban system to provide sufficient livelihood for everyone. The consequences are that the most vulnerable part of the society – urban poor – are possibly most challenged by the lack of basic services and impacts of climate change as well, according to Hardoy and Pandiella (2009).

Urban Water Supply

Another emerging risk due to rapid urbanisation is widely seen in exhausted basic services (Satterthwaite et al., 2007; UNHABITAT, 2008, 2010; UNISDR, 2009). This means that the provisions of services, for example, water, electricity or sanitation, and management of solid waste disposal is insufficient and thus emerge as risk factors that may either support the occurrence of a disaster or hamper the recovery process from such an event.

Fig. 3 shows the percentage of people that have access to freshwater in urban areas (UNESCAP, 2011). Despite the fact that residents in urban areas have higher access to water sources, it does not necessarily mean that an access rate of more than 90 per cent in all regions in Asia implies no problems in the actual supply.

In contrast, according to McDonald et al. (2011), the number of urban people in Asia going to be directly affected by perennial and non-perennial (require short-scale investments to transport water within a radius of <100 km) water shortage will dramatically increase from 219 million people in 2000 to 882 million people in 2050. In comparative terms, 19.7 per cent of the urban population in Asia was affected by water shortage which is estimated to increase to 23.4 per cent in 2050.

The equivalent figures for Africa are 33.5 per cent in 2000 and 44 per cent in 2050. Fig. 4 shows the spatial distribution water shortages in cities (population >1 million) in Latin America, Africa and Asia. In particular, urban areas in the South Asian and Eastern part of China are expected to be severely challenged seasonal water shortage. The trigger for this predicted water shortage in many urban parts of the world is due to impacts of urbanisation and climate change which will require massive investments into water-related infrastructures (McDonald et al., 2011).

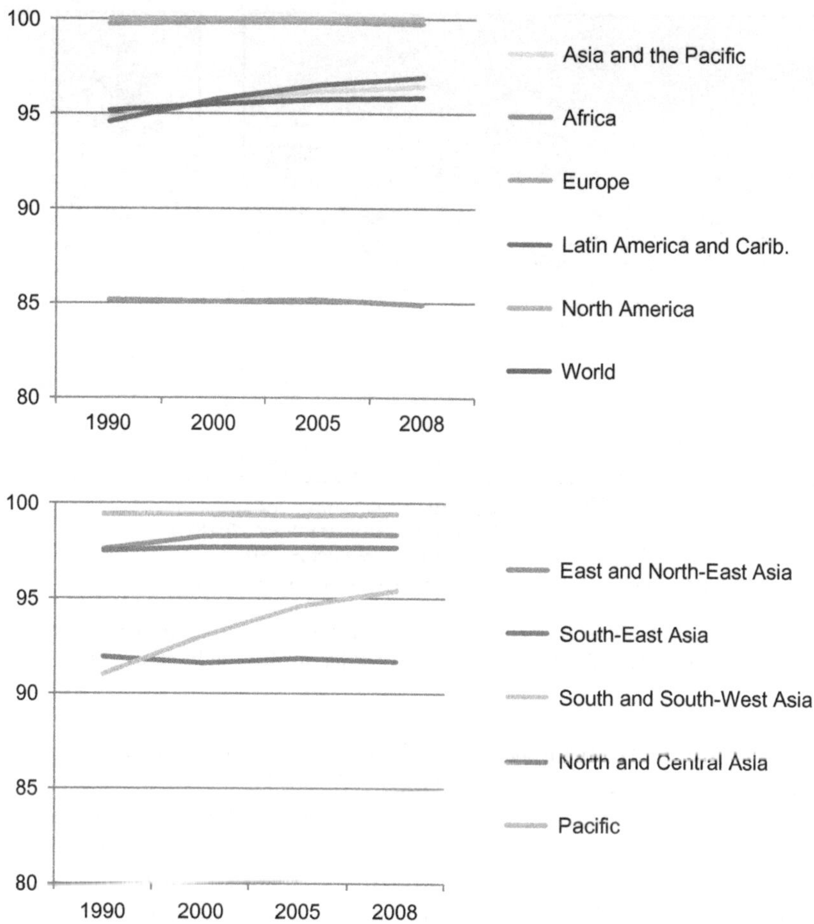

Fig. 3. Access to Water Sources in Urban Areas in the World (Top) and Asia (Bottom) between 1990 and 2008. *Source*: Modified from UNESCAP (2011).

In the case of India, the share of population who has water connection at their houses is low in India's urban areas with only 64 per cent compared to urban areas in China (91 per cent), South Africa (86 per cent) or Brazil (80 per cent) (GoIMUD, 2011). Furthermore, the daily duration of water supply is only provided within a range of 1–6 hours compared to 24 hours in China and Brazil. Finally, the water network in urban areas of India is challenged by neglected maintenance which is causing leakages and loss of

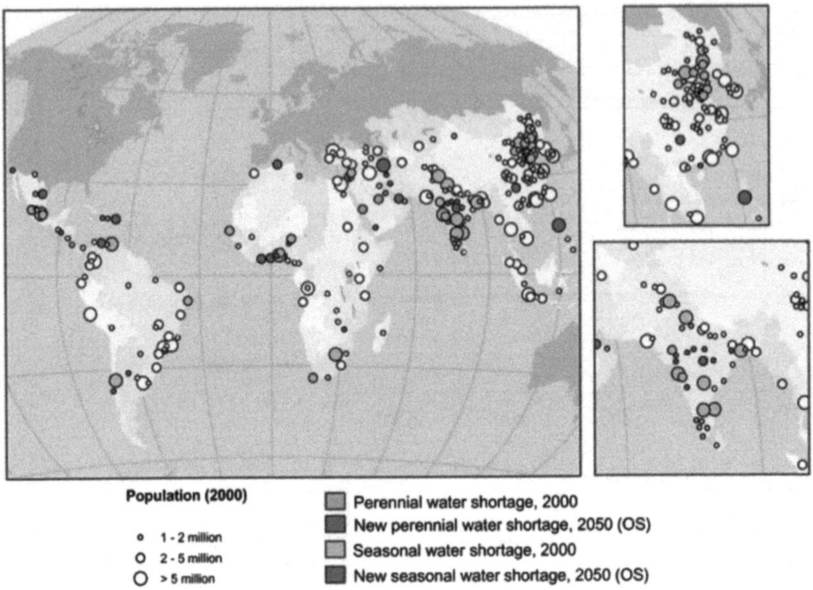

Fig. 4. Spatial Distribution of Large Cities (>1 million) and Their Water Shortage Status, in 2000 and 2050. *Source*: Adapted from McDonald et al. (2011).

available water. Accordingly, it is estimated that 70 per cent of the pipes connecting consumers are leaking (GoIMUD, 2011).

The ongoing urbanisation of numerous Indian cities compounds the water supply problem which often leads to enhanced extraction of groundwater. This is visible, for example, in various Indian cities, such as Jaipur, Chennai and Delhi. The consequences of groundwater loss due to excessive extraction (a result of increasing consumption due to urbanisation) can lead to interrupted daily services and land subsidence. In order to alleviate these pressures, large-scale measures are often sought to overcome this problem. An ideal example is Jaipur, located in Rajasthan (relatively dry, average 668 mm rain per year), where a pipeline to the 120 km far away Bisalpur Dam was built during the 1990s to supply the city with water (Government of Rajasthan (GoR), 2004). Before the supply of water from this dam started in 1998, 97 per cent of water supply was extracted from Jaipur's groundwater causing severe instability of the soil structure and land subsidence (GoR, 2004). This example demonstrates a type of emerging risk due rapid urban growth in Jaipur. In 2011, the city had 3.7 million inhabitants with an average yearly growth rate in of 4.4 per cent during the period from 1981 to 2011. This Bisalpur project, co-financed by international donors such as the Japan International Cooperation Agency (JICA) and Asian Development Bank (ADB), demonstrates further that without external support in form of far located water sources and international financial support, Jaipur itself would not have been able to absorb the impacts of urbanisation. While the first phase (completed) of this project is expected to provide water to 2.2 million people, the second phase is planned to be completed by 2021 and targets the supply for a total 5 million people. However, until today people suffer from water shortage, particularly if the monsoon period is delayed in July.

This sudden dependence on external water supply is a serious risk driver for disasters and has been taken-up by the Government of Rajasthan to impose a State Water Policy in 2010 to urge not only urban dwellers but also residents in rural areas to conserve the use of water (GoR, 2010). Although the Government of India may welcome this initiative by the Government of Rajasthan, it also advocates for enhanced efforts to tackle the issue of water scarcity by supporting local water recycling (treatment) solutions instead of transporting water over long distances simply because of higher cost-efficiency (Government of India, 2011).

Urban Electricity Supply

While the provision of water supply is a challenge for many cities, particularly in developing countries, the provision of electricity is another major challenge for cities and emerges as a risk driver due to urbanisation. Although, in general, the provision of electricity is better in urban than rural areas of a country (National Research Council, 2003), the sudden increase of people, especially in small cities, can overwhelm the existing capacities (Cohen, 2006). As a result, electricity is usually provided in cities, but is impaired by irregular and interrupted services. The reasons for these power outages are, however, not only due a lack of energy capacities but much more because of poor maintenance of the electricity grid and illegal connections.

Urban Sanitation and Solid Waste Management

As more people come to a city, more waste is produced. Unfortunately, urban areas in developing countries are often challenged to meet the rising demands of proper sanitation and waste management services (UNHABITAT, 2008; UNISDR, 2009). Fig. 4 shows the percentage of people in urban areas who have access to improved sanitation (UNESCAP, 2011). Hence, urban populations in Africa and Asia are having in average lower access to sanitation compared to the world average. In particular, regions in South and South-West Asia are having lowest rates around 60 per cent. Although, one has to be careful in jumping into the conclusion that the access to sanitation would be dramatically low in urban areas in Africa and Asia; however, urban areas are still providing more improved sanitation access to people compared to rural areas (UNESCAP, 2011). Thus, the comparatively low access rates of 51 per cent in Nepal, 54 per cent in India or 58 per cent in China, of people in urban areas, do not highlight that significantly less people are having the equal access in rural areas (27 per cent in Nepal, 21 per cent in India and 52 per cent in China). However, since the urban population is absorbing most of the current population growth access rate ought to increase to enhance the living quality in cities and provision of a healthy society (Fig. 5).

In the case of India, a study undertaken by the Ministry of Urban Development concluded that in 4,861 out of 5,161 cities and towns in India, the urban areas do not even have partial sewerage network (GoIMUD, 2010).

Urban solid waste management is another serious risk driver which is due to a lack of solid waste garbage collection in urban areas. Accordingly, only 30–60 per cent of solid waste is collected in urban areas in developing countries in Asia and is instead dumped into streets, drains, rivers, etc. (ADB, 2008). According to the Asian Productivity Organization (APO) (2007), only an amount of 0.15 kg of solid waste is produced in rural areas per person and per day which stands in stark contrast to urban populations which have much higher waste production that can be as high as 1 kg per day and per person (Table 2).

However, the high percentage of organic matter in the solid waste in Asian urban areas between 50 and 70 per cent still allows waste to be destroyed naturally after some time (APO, 2007). Nonetheless, the rising production of waste is a serious issue threatening the health and livelihood of people (ADB, 2008).

In the case of India, the amount of produced solid waste is still much lower in urban areas compared to other cities, such as Bangkok, Taipei, etc. However, the management (segregation, recycling, etc.) and collection is a major challenge (GoIMUD, 2011). Thus, solid waste may not be collected and destroyed appropriately. This is also to some extent due to the high costs imposed on Municipalities which often causes a financial burden; accordingly, 25–50 per cent of an Indian Municipalities' budget may account for solid waste disposal (World Bank, 2008).

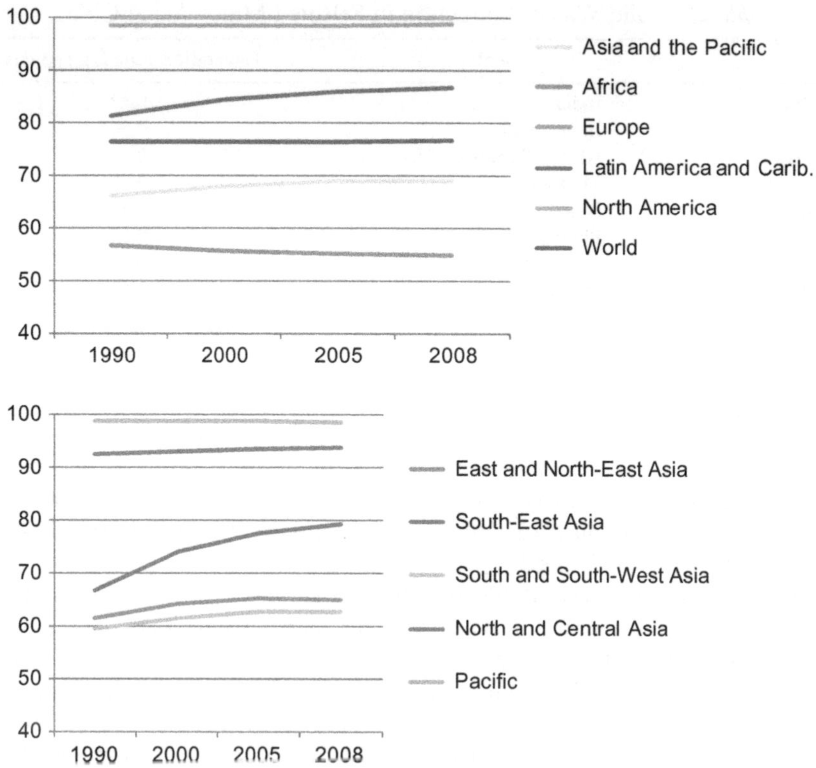

Fig. 5. Access to Improved Sanitation in Urban Areas in the World (Top) and Asia (Bottom) between 1990 and 2008. *Source*: Modified from UNESCAP (2011).

Built Environment and Loss of Urban Greenspace

Another emerging risk due to urbanisation is the rising demand on housing and available land (UNISDR, 2009). The demand is two-fold: on one hand more people due to migration trends urban areas also mean higher requirements, but on the other changing living habits demand more spacious living condition and thus require more living floor space in an already densely populated area. The consequences of these rising demands are that more land is needed in particular for housing purposes. Looking at India, for example, the absence of implemented housing strategies led in most cities to an urban sprawl trend, where the urban areas are growing much more vertically than horizontally. Therefore, the logic consequence is that agricultural land and also other greenspace (forests, parks, etc.) has to serve

Table 2. Solid Waste Production in Selected Major Asian Cities.

City	Country	Generation Rate (kg/cap/day)
Delhi	India	0.47
Dhaka	Bangladesh	0.50
Urban	Islamic Republic of Iran	0.80
Penang	Malaysia	0.98
Kathmandu	Nepal	0.30
Manila	Philippines	0.66
Singapore	Singapore	0.94
Colombo	Sri Lanka	0.62
Taipei	Republic of China	0.95
Bangkok	Thailand	0.88
Hanoi	Vietnam	0.63

Source: Adapted from APO (2007).

for new development areas. This is also the case in Chennai, Hyderabad and other major Indian cities which have experienced the expansion of their urban areas at a rapid speed.

Indian cities are unavoidable and this is a serious problem reducing the quality of life and increasing the risk of disasters. Accordingly, less urban greenspace reduces the absorption capacity of a city following heavy rainfall events (Gill, Handley, Ennos, & Pauleit, 2007). Hence, surface run-off of water is becoming higher which can quickly lead to water-logged areas and other types of floods. Furthermore, the loss of urban greenspace adds to the urban heat island effect, thus increasing daily peak temperatures.

As a result, the loss of greenspace in urban areas of developing countries is taking in many cases the same trajectory as already seen in cities in developed countries, such as Tokyo or Seoul. However, Tokyo also serves as a good example where the loss of greenspace has been recognised and new strategies have been implemented by the Metropolitan Government to regain the loss of greenspace through the promotion of green roofs, plantation of trees in streets, enhancement of parks, etc. (Tokyo Metropolitan Government, 2007).

Disrupted Ecosystems

Although the loss of greenspace and agricultural land in Asian urban areas has been replaced by new built environments, the consequences are the loss and extinction of natural habitats for the flora and fauna (United Nations Environment Programme (UNEP), 2002, 2007). In many cases, however, the impacts of urbanisation and associated economic growth do only disrupt the ecosystem of cities. In particular, burning of fossil fuel has severe detrimental impacts on urban areas' air quality. Fig. 6 exemplifies the above; the World Health Organization (WHO) tolerated levels of fine dust (PM10) in major Asian cities. For example, Beijing and Xian have levels which are six times higher than the WHO norm (Clear Air Initiative for Asian Cities (CAI-Asia), 2006). Although, development may lead to a more intense use of natural resources and ecosystems, fine dust levels in Singapore, highly developed Asian city, are just a little above the tolerated WHO norm. This makes clear that urban development does not automatically cause cities to become hot-spot for environmental degradation. Nevertheless, not only many cities in China and India but also other Asian countries are challenged by severe pollution of their urban water bodies and soil contamination. For instance, in Indian cities only 27 per cent of urban household waste water gets treated (GoIMUD, 2011). Another, unfortunately, negative examples are Hanoi and Ho Chi Minh City where as of 2008, residential, hospital and industrial waste water were almost not

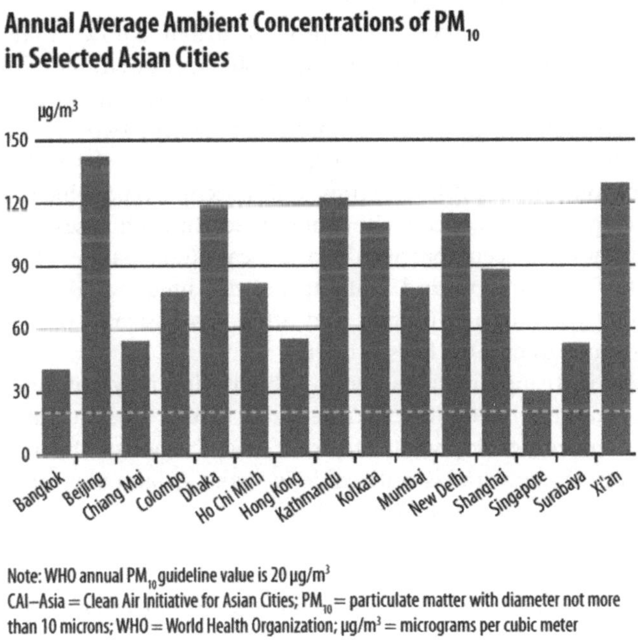

Fig. 6. Fine Dust Levels in Selected Asian Cities. *Source*: Adapted from CAI-Asia (2006).

treated at all before discharged into the nearby river streams (ADB, 2008). Accordingly, more than 70 per cent of industrial zones and 90 per cent of production units do not have adequate treatment facilities.

However, there is also a good example where polluted water bodies are cleaned-up and waste water becomes treated. In the 'Three River Master Plan' in Metro Manila, the cleaning of the Marikina, San Juan and Pasig River was made official. The plan was adopted in 2010.

Exhausted Institutions and Unplanned Development

The term exhausted institutions aims to explain the often dramatic challenges which are faced not only by local urban governments in Asia but also in other developing countries, to cope with rapid changes in their urban areas. Thus, city governments may find it difficult to enforce laws regarding environmental pollution, housing and denial of planning projects that would cause harm to the environment. As a result, environmental problems are created due to poor institutional management which does, however, not necessarily stand in direct relationship to pressure arising from urbanisation (UNEP, 2002).

Although there are many examples of urban sprawl that inflicted unplanned development, Li and Wang (2008) reviewed the situation in China, where the illegal claiming of land is a serious issue in urban fringe areas. Accordingly, Haikou and Nanjing experiences two types of illegal constructions which are due to self-help housing which refer to houses that are built beyond the legal boundaries and a second type which is the real estate development on collectively owned land. In both cases, the acquisition of land has not been approved by the local municipalities (Fig. 7).

The case from Haikou and Nanjing are not unique and take place in many rapidly growing urban areas and often demonstrates the inability of local institutions to enforce laws properly.

To conclude, rising pressures due to urbanisation may not directly cause a local institution to become exhausted and neither does it automatically lead to unplanned development. However, the absence of strong functioning regulatory frameworks is noticeable in many Asian cities, also in India (National Institute of Urban Affairs, 2005).

Pressure from Impacts of Climate Change in Urban Areas

According to the projections published in the Fourth Assessment Report by the IPCC, it is expected that the climate is likely to change in the future

Fig. 7. Illegal Constructions in Haikou (Left) and Nanjing (Right). *Source*: Adapted from Li and Wang (2008).

(IPCC, 2007). As a result, weather is projected to become less predictable and thus extreme events may become more frequent and intense in the coming decades. The projected temperature increases are varying in Asia from 3 to 5 degree Celsius during the 21st century compared to the year 2000. In particular, northern areas in Asia will experience higher average temperatures.

In the case of India, it is projected that annual mean temperatures will rise between 3.5 to 5 and 2.5 to 4 degree Celsius during the 21st century depending on the simulation model (Kumar et al., 2006). These temperature increases are likely to become amplified in urban areas due the urban metabolism which adds to higher daily maximum temperatures (urban heat island effect) (Kovats & Akhtar, 2008).

Although it is expected that climate-related hazards become more frequent and more severe at global scale (IPCC, 2007), this trend cannot be confirmed so far for Indian cities, according to studies from De, Dube, and Prakasa Rao (2005), where they cannot confirm increases of natural hazards in the past century due to climate change rather than a decrease of numbers of cyclones in the Bay of Bengal. While these observations on past data, Sathaye, Shukla, and Ravindranath (2006) predict that there is a 20 per cent rise in summer monsoon rainfall (July–September) for most parts of India, except for states, like Rajasthan, Punjab and Tamil Nadu.

However, regardless of whether climate change is already causing more severe and frequent natural hazards, urban areas are at risk to become hot-spots for hydro-meteorological disasters, as the following examples will point out. This assumption is done despite the absence of available disaster records for urban areas (Fig. 2 only shows the overall increase of disasters in Asia and the world) that would prove that cities are increasingly affected by disasters. Adopting an event-based type of analysis of recent events aims to compensate this shortcoming. The challenge of justifying the increasing disaster events in urban areas is also acknowledged by the International Federation of Red Cross and Red Crescent Societies (IFRCRCS) in their latest World Disaster Report 2010 which focused on urban risk.

Urban Flooding

The most recent flooding disaster in Bangkok in October 2011 exemplified the devastating impacts of above average amounts of monsoon rainfall. Thus, the flooding stretched over wide parts of the north of Thailand until and began in July 2011 until water reached the centre of Bangkok in the middle of October, which is located in flat area with little capacity to absorb water due to its low altitude, just a few metres above sea level. The total death toll has exceeded 600 on 20 November 2011 (Associated Press, 2011). Although few people were killed in Bangkok, the city is still in many parts affected by water-logging even after six weeks since the water reached the centre of the city. Estimations of economic costs are currently being investigated.

Another extreme urban flooding event was in Mumbai in 2005 where around 1,000 people died due to heavy rainfalls (Revi, 2008). Within 24 hours 944 mm of rainfall were recorded and caused immediate flooding (Kovats & Akhtar, 2008). This disaster also demonstrated the limited availability of proper urban drainage systems in Mumbai, which is likewise the case for many other cities in India (Kovats & Akhtar, 2008). Moreover, the lack of catchment areas for large rainfall quantities, leading to flooding, may cause contamination of other water bodies with chemicals and other hazardous substances which again lead to potential health problems of citizens. This is a particular problem in urban areas where slums are often located near rivers and canals. Indirect impacts of climate change, like sea-level rise, is an additional stress for low-lying urban areas, like Chennai, Calcutta or Mumbai, which may lead to salt water intrusion into groundwater bodies (Revi, 2008).

Mulyasari, Shaw, and Takeuchi (2011) reviewed past urban flood events which were caused by intense rainfall and caused casualties and significant economic damages to the urban infrastructures. Hence, various cities across Asia were affected by urban floods during the past decades which are as follows: Nagoya in 2000 (567 mm precipitation, 10 casualties), Taipei in 2001 (530 mm precipitation, 104 casualties), Mumbai in 2005 (as mentioned earlier), Jakarta in 2007 (750 mm precipitation, 80 casualties) and Hanoi in 2008 (450 mm precipitation, 85 casualties). Although this list can easily be enhanced, it highlights that most Asian countries and particularly cities close to rivers and coastal areas are prone to flooding events. In this context, Metro-Manila needs to be mentioned which is possibly one of the most flood-prone urban area in Asia with up to 18 flood events every year (Zoleta-Nantes, 2000).

Sea-Level Rise

Looking particularly at Asia and its urban areas, it is expected that 'especially heavily populated megadelta regions in South, East and South-East Asia, will be at greatest risk due to increased flooding from the sea and, in some megadeltas, flooding from the rivers' (IPCC, 2007, p. 50). Thus, sea-level rise is expected to be a major challenge for many coastal cities which are located at low elevation (McGranahan, Balk, & Anderson, 2007). According to the study of McGranahan et al. (2007), around 360 million urban dwellers (globally) lived in low-elevation coastal zones in 2000 which represent around 57 per cent of the people located in these areas. Additionally, 9 out of 10 countries in the top 10 list of population living in the Low Elevation Coastal Zone (LECZ) are from the Asian continent. The country with the highest population in the LECZ is China which has many newly urbanising areas located pre-dominantly in coastal areas which are exposing large numbers of people to potential impacts of climate change not only in the form of sea-level rise but also less predictable storms causing direct wind damages and flooding.

In the case of India, various large cities and also so-called megacities are either directly located along the 5,700 km long (mainland) Indian coastline or in its vicinity, such as Chennai, Mumbai or Calcutta, apart from other numerous small- and medium-sized cities (Aggarwal & Lal, 2001; Revi, 2008). According to the estimations of Aggarwal and Lal (2001), sea-level rise in India in the coming decades will be around 2.5 mm per year compared to 1 mm observed in the past. This increase of sea-level rise 'speed' is not only due to ice losses in Antarctica and Greenland but also due to spatial expansion of warmer water (IPCC, 2007). Thus, it is expected that the sea waters around India's coasts are going to rise between 46 and 59 cm during the 21st century (Aggarwal & Lal, 2001). As a result, people living in low-lying areas, such as Chennai, are exposed to slow but steady sea-level rise. Table 3 shows that around 63 million (or around 5 per cent of

Table 3. Ranking of People Exposed to Sea-Level Rise (Globally).

Top Ten	Ranked by Total Population in the LECZ				Ranked by Share of Population in the LECZ			
	Country	Overall rank[a]	Population in the LECZ		Country[b]	Overall rank[a]	Population in the LECZ	
			Counts ('000)	%			Counts ('000)	%
1	China	1	143,880	11	Bahamas	174	267	88
2	India	2	63,188	6	Suriname	170	318	76
3	Bangladesh	8	62,524	46	Netherlands	59	11,717	74
4	Vietnam	13	43,051	55	Vietnam	13	43,051	55
5	Indonesia	4	41,610	20	Guyana	157	415	55
6	Japan	9	30,477	24	Bangladesh	8	62,524	46
7	Egypt	16	25,655	38	Djibouti	160	289	41
8	USA	3	22,859	8	Belize	179	91	40
9	Thailand	19	16,478	26	Egypt	16	25,695	38
10	Philippines	14	13,329	18	The Gambia	150	494	38

Source: Adapted from McGranahan et al. (2007).
[a]Refers to overall rank in total population.
[b]Countries with a total population of under 100,000 people, or smaller than 1,000 km^2, were excluded from this list; if all countries were included, 7 of the top 10 would be places with fewer than 100,000 persons, the top 5 having more than 90 per cent of their country in the LECZ (Maldives, Marshall Islands, Tuvalu, Cayman Island, Turk and Caicos Island).

the total population) of people living in the LECZ, people located within 10 m above sea level, in India are estimated to be affected by sea-level rise. Thereby, the most vulnerable states are expected to be Gujarat and Tamil Nadu (Janakarajan, 2007).

Rainfall-Induced Landslides

Petley et al. (2007) assessed the urban landslides in Nepal and raised the issue that the risk of landslide is often underestimated in planning aspects. Hence, cities located along the southern part of the Himalayan range are at particular risk to experience landslides after rainfall events. For example, Shimla, located in the north of India and increasing urban population, is built at a hilly area with slopes as steep as 60 degrees. Not surprisingly, this city with a population of more than 400,000 is frequently affected by rainfall-induced landslides, usually every year smaller or bigger events occur. Chapter 5 discusses the resilience of Shimla, based on the CDRI results. Other countries affected by rainfall-induced landslides are most countries in Asia with a hilly landscape.

Storms

Many parts of Asia are severely prone to cyclone-related storm events. During the monsoon period from June to November (depending on the location) storms establish over the Bay of Bengal and the Pacific which then take landfall in the northerly located countries. Similar to sea-level rise and drought, it may not be possible to localise storms specifically to only a single urban entity. However, the impacts felt in urban areas may be of dramatic consequences. While high wind speeds may cause initial damage to people and infrastructure, in many cases these storms or cyclones are also bringing heavy amounts of rainfall. For example, in 2008, Cyclone Nargis devastated large parts of Myanmar including the city of Yangon. This event caused 138,366 casualties and a damage amount of USD 4 billion (IFRCRCS, 2011). Although this damage amount may look small compared to the costs due to Hurricane Katrina in 2005 that stood at USD 125 billion, it exemplifies that in many developing countries in Asia (including urban areas), the costs on human lives is often higher than in terms of infrastructural damage (UNISDR, 2009). Apart from Myanmar, the Philippines, Japan, Taiwan, China, Vietnam, Bangladesh and India are the countries which are most susceptible to storms, usually in the form of cyclones. Of greater concern is, particularly, China's growing urban coastal population which is directly exposed to storms.

Drought and Heat Waves

While sea-level rise is to some extent predictable (IPCC, 2007; McGranahan et al., 2007), the yearly occurrence of droughts and heat waves are only partially projectable.

However, it is expected that heat waves and droughts will increase in numbers and result in losses of human lives and rising health problems of people (Kovats & Akhtar, 2008; Revi, 2008). Unlike heat waves, droughts have a slow-onset which allows more time for preparedness measures, although their exact occurrence and intensity (length and force) is expected to change in the future (IPCC, 2007). As a result, changing climate patterns are expected to have serious impacts on the supply of water in drought and heat-wave-prone urban areas. Furthermore, stress due to absence of rainfall leads to rural–urban migration trends which are an additional compounding factor to increase the demand of water in urban areas (Van der Bruggen, Borghgraef, & Vinckier, 2010).

In the case of India's cities, the western parts of India and areas along the Ganges are most vulnerable to increased occurrence of droughts (Janakarajan, 2007).

Linkage between Urban Risk and Impacts of Climate Change

The pervious sections have emphasised on different emerging urban risk – so-called risk drivers. To facilitate the understanding, these risk drivers shall be termed as stresses for the optimal functioning of a city which are long-lasting and enduring over time. In contrast, impacts from climate-related hazards are termed as shocks due to their often unexpected occurrence. Fig. 8 shows graphically the relationship between stresses and shocks.

The theoretical understanding between stresses and shocks, shown in Fig. 3, is closely orientated to the 'Pressure and Release (PAR)' model of Wisner, Blaikie, Cannon, and Davis (2004) which points out that a higher intensity or number of pressures, in our terminology stresses, may overwhelm a defined system, such as a city, if additional unexpected pressure (shock) is added to it.

The linkage between stresses and shocks leading to disaster is based on the notion that a disaster is likely to happen if various systems (social, economic, physical, institutional and environmental) within a community fail to cope with a specific natural hazards (Comfort et al., 1999; Hewitt, 1997). In other words, in the context of urban areas climate disaster risk may exacerbate and worsen the impact of a natural hazard.

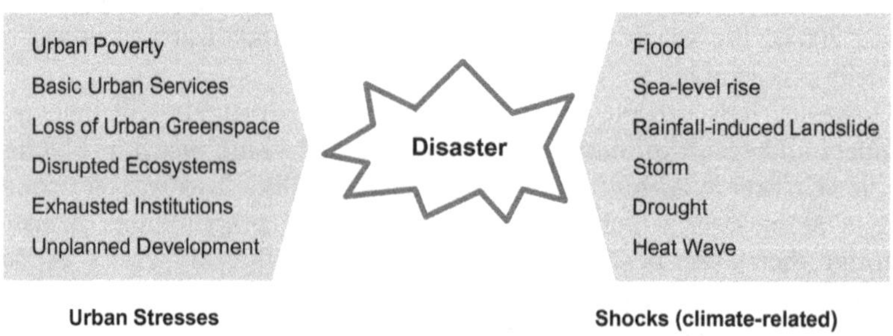

Fig. 8. Stresses and Shocks in an Urban Area.

REFERENCES

Aggarwal, D., & Lal, M. (2001). *Vulnerability of Indian coastline to sea-level rise*. New Delhi: Centre for Atmospheric Sciences, Indian Institute of Technology.

Asian Development Bank (ADB). (2008). *Toward resource-efficient economies in Asia and the Pacific*. Manila: Institute of Global Environmental Strategies ADB.

Asian Productivity Organization (APO). (2007). *Solid waste management. Issues and challenges in Asia*. Tokyo: APO.

Associated Press (AP). (2011, November 20). *Death toll from Thailand's floods tops 600*, AP. Retrieved from http://www.google.com/hostednews/ap/article/ALeqM5jfpkevwPmB4kvdXgjOu0O-ZTt_yg?docId=f74c7f649ed848a787b0bfee140b57c8. Accessed on December 3.

Census of India. (2011). *Provisional population totals. Government of India, paper 2, vol. 1 of 2011 rural-urban distribution, India series 1*. Retrieved from http://www.censusindia.gov.in/2011-prov-results/paper2/data_files/india/paper2_1.pdf. Accessed on October 30.

Clear Air Initiative for Asian Cities (CAI-Asia). (2006). *Urban air quality management. Summary of country/city synthesis reports across Asia*. Asian Development Bank, Discussion Draft. Retrieved from http://cleanairinitiative.org/portal/sites/default/files/documents/overview_0.pdf. Accessed on November 11, 2011.

Cohen, B. (2006). Urbanization in developing countries: Current trends, future projections, and key challenges for sustainability. *Technology in Society*, *28*, 63–80.

Comfort, L., Wisner, B., Cutter, S., Pulwarty, R., Hewitt, K., Oliver-Smith, A., … Krimgold, F. (1999). Reframing disaster policy: The global evolution of vulnerable communities. *Environmental Hazards*, *1*, 39–44.

De, U. S., Dube, R. K., & Prakasa Rao, G. S. (2005). Extreme weather events over India in the last 100 years. *Journal of the Indian Geophysical Union*, *9*(3), 173–187.

Gill, S. E., Handley, J. F., Ennos, A. R., & Pauleit, S. (2007). Adapting cities for climate change: The role of green infrastructure. *Built Environment*, *33*(1), 115–133.

Government of India (GoI). (2011). *Draft – Faster, sustainable and more inclusive growth. An approach to the 12th five year plan*. New Delhi: GoIPC.

Government of India Ministry of Urban Development (GoIMUD). (2010). *National rating and award scheme for sanitation for Indian cities*. New Delhi: GoIMUD.

Government of India Ministry of Urban Development (GoIMUD). (2011). *Report on Indian urban infrastructure and services*. Retrieved from http://niua.org/projects/hpec/FinalReport-hpec.pdf. Accessed on November 4.

Government of India Planning Commission (GoIPC). (2009). *Report of the expert group to review the methodology for estimation of poverty*. New Delhi: GoIPC.

Government of Rajasthan (GoR). (2004). *Full resettlement plan, Bisalpur water supply project, Rajasthan urban infrastructure development project (ADB loan no. 1647-IND)*. Government of Rajasthan Urban Development Department, Jaipur. Retrieved from http://www.adb.org/Documents/Resettlement_Plans/IND/BisalpurWater/bisalpur-water.pdf. Accessed on October 20, 2011.

Government of Rajasthan (GoR). (2010). *State water policy*. Jaipur: Government of Rajasthan State Water Resource Planning Department.

Hardoy, J., & Pandiella, G. (2009). Urban poverty and vulnerability to climate change in Latin America. *Environment and Urbanization*, *21*, 203–224.

Hewitt, K. (1997). *Regions of risk: A geographical introduction to disasters*. Essex: Longman.

IBN-Live. (2009). *Earn Rs 32 a day? Not poor enough.* Retrieved from http://ibnlive.in.com/news/earn-rs-32-a-day-not-bpl-planning-commission/186003-3.html. Accessed on September 21, 2011.

International Federation of the Red Cross and Red Crescent Societies (IFRCRCS). (2011). *World disaster report 2010, focus on urban risk.* Geneva: IFRCRCS.

Intergovernmental Panel on Climate Change (IPCC). (2007). Summary for policymakers. In S. Solomon, D. Qin, M. Manning, Z. Chen, M. Marquis, K. B. Averyt, ... H. L. Miller (Eds.), *Climate change 2007: The physical science basis. Contribution of Working Group I to the Fourth Assessment Report of the Intergovernmental Panel on Climate Change.* Cambridge, MA: Cambridge University Press.

Janakarajan, S. (2007). Challenges and prospects for adaptation: Climate and disaster risk reduction in coastal Tamil Nadu. In M. Moench & A. Dixit (Eds.), *Working with the winds of change* (pp. 235–270). Boulder, CO: ISET.

Kovats, S., & Akhtar, R. (2008). Climate, climate change and human health in Asian cities. *Environment and Urbanization, 20*, 165–175.

Kumar, R. K., Sahai, A. K., Krishna Kumar, K., Patwardhan, S. K., Mishra, P. K., Revadekar, J. V., ... Pant, G. B. (2006). High-resolution climate change scenarios for India for the 21st century. *Current Science, 90*(3), 334–345.

Li, Z., & Wang, W. (2008). *Illegal construction on the urban fringe as new landscape of urban sprawl: The case of Nanjing, China.* 44th ISOCARP Congress, 2008. Retrieved from http://www.isocarp.net/Data/case_studies/1259.pdf. Accessed on November 20, 2011.

McDonald, R. I., Green, P., Balk, D., Fekete, B. M., Revenga, C., Todd, M., & Montgomery, M. (2011). Urban growth, climate change, and freshwater availability. *Proceedings of the National Academy of sciences of the United States of America* (pp. 1–6). New York, NY.

McGranahan, G., Balk, D., & Anderson, B. (2007). The rising tide: Assessing the risks of climate change and human settlements in low elevation coastal zones. *Environment and Urbanization, 19*(1), 17–37.

Mulyasari, F., Shaw, R., & Takeuchi, Y. (2011). Urban flood risk communication for cities. *Community, Environment and Disaster Risk Management, 6*, 225–259.

National Institute of Urban Affairs (NIUA). (2005). *Impact of the constitution (74th Amendment) act on the working of urban local bodies* (vol. 1, Series 106). Retrieved from http://www.niua.org/research_studies_2006.asp. Accessed on October 19, 2011.

National Research Council (NRC). (2003). *Cities transformed: Demographic change and its implications in the developing world.* Washington, DC: National Academies Press.

Pelling, M. (2003). *The vulnerability of cities: Natural disasters and social resilience.* London: Earthscan.

Petley, D. N., Hearn, G. J., Hart, A., Rosser, N. J., Dunning, S. T., Oven, K., & Mitchell, W. A. (2007). Trends in landslide occurrence in Nepal. *Natural Hazards, 43*(1), 23–44.

Revi, A. (2008). Climate change risk: An adaptation and mitigation agenda for Indian cities. *Environment and Urbanization, 20*(1), 207–229.

Sathaye, J., Shukla, P. R., & Ravindranath, N. H. (2006). Climate change, sustainable development and India: Global and national concerns. *Current Science, 90*(3), 314–325.

Satterthwaite, D., Huq, S., Pelling, M., Reid, H., & Romero-Lankao, P. (2007). *Adapting to climate change in urban areas: The possibilities and constraints in low- and middle-income countries.* Human Settlements Climate Change and Cities Discussion Series 1. London: IIED.

Silva, T. (1996). Poverty and uneven development: Reflections from a street children project in the Philippines. *Childhood, 3*, 279–282.

Tokyo Metropolitan Government (TMG). (2007). *Basic policies for the 10-year project for green Tokyo – Regenerating Tokyo's abundant greenery*. Retrieved from http://www.kankyo.metro.tokyo.jp/en/attachement/10-year_project.pdf. Accessed on October 20, 2011.

United Nations Economic and Social Communication for Asia and the Pacific (UNESCAP). (2011). *Statistical Yearbook for Asia and the Pacific 2011*. Bangkok: United Nations.

United Nations Environment Programme (UNEP). (2002). *Global Environmental Outlook 3. Past present and future perspectives*. London: Earthscan.

United Nations Environment Programme (UNEP). (2007). *Global Environmental Outlook 4. Environment for development*. Valletta: Progress Press Ltd.

United Nations Human Settlements Programme (UNHABITAT). (2003). *The challenge of slums: Global report on human settlements*. London: Earthscan.

United Nations Human Settlements Programme (UNHABITAT). (2008). *State of the world's cities 2008/2009*. London: Earthscan.

United Nations Human Settlements Programme (UNHABITAT). (2010). *The state of Asian cities 2010/11*. London: Earthscan.

United Nations International Strategy for Disaster Reduction (UNISDR). (2009). *Global Assessment Report on Disaster Risk Reduction*. Geneva: United Nations.

United Nations Population Division (UNPD). (2010). *World urbanization prospects: The 2009 Revision*. New York, NY: United Nations Population Division, Department of Economic and Social Affairs United Nations.

Van der Bruggen, B., Borghgraef, K., & Vinckier, C. (2010). Causes of water supply problems in urbanised regions in developing countries. *Water Resource Management, 24*, 1885–1902.

Wisner, B., Blaikie, P., Cannon, T., & Davis, I. (2004). *At risk: Natural hazards. People's vulnerability and disasters*. Cambridge, MA: MIT Press.

World Bank. (2008). *Improving management of municipal solid waste in India: Overview and challenges*. Washington, DC: World Bank.

Wratten, E. (1995). Conceptualizing urban poverty. *Environment and Urbanization, 7*, 11–38.

Zoleta-Nantes, D. B. (2000). Flood hazards in Metro Manila: Recognizing commonalities, differences and courses of action. *Social Science Diliman, 1*(1), 60–105.

CHAPTER 3

THE CONCEPT OF RESILIENCE TO DISASTERS

THE CONCEPT OF RESILIENCE

The previous sections have emphasised on the various risks emerging in urban areas, particularly in developing countries, due to rapid inflow of people and impacts of climate change. In this section, the aim is to lay the methodological foundation by using the concept of resilience to later assess the resilience of urban areas to climate-related disasters. Therefore, the objective is to transform the concept in a way that it can describe processes in cities as well as urban neighbourhoods by focusing on different aspects of the functioning of a community in the advent and during a climate-related disaster.

At first, this section defines the term resilience and its elements. Secondly, the concept of resilience is discussed in relation to other concepts, like vulnerability, before gradually defining how the disaster resilience of urban areas and their communities can be defined through the concept of resilience.

Defining Resilience

One of the first approaches to describe resilience has its origin from the field of ecology where Holling (1973, p. 14) describes it as 'a measure of the persistence of systems and of their ability to absorb change and disturbance and still maintain the same relationships between populations or state variables'. Transforming this concept into the urban disaster setting that would mean simply whether an urban area can absorb a disaster (disturbance, shock) or not. However, before jumping into preliminary conclusions, the origin of the previous thought has its grounds from theoretical evolutions of the concept of resilience to include social aspects alongside of ecological, which needs further explanations. Thus, Adger (2000) draws a linkage on how human beings respond to social, political and environmental changes. In his work, he emphasises on the abilities of communities to secure their livelihood while absorbing impacts of different types of external changes of their living environment. As a result, an interdisciplinary connection is drawn to reflect and address the complex environment in which communities are embedded, in order to determine their resilience.

This linkage drawn by Adger (2000) points out that the 'populations or state variables' described by Holling (1973, p. 14) may also refer to human beings rather than only animal populations. As a result, Adger (2000) unveiled the term social resilience or socio-ecological resilience to describe the interconnections between human beings and disruptions in different environments, ranging from social, economic, political and natural.

Following these arguments, Carpenter, Walker, Anderies, and Abel (2001) evolve on this theoretical understanding by interpreting socio-ecological resilience as a system which has the following properties:

- 'the amount of change the system can undergo and still remain within the same domain of attraction;
- the degree to which the system is capable of self-organization;
- and the degree to which the system can build the capacity to learn and adapt' (Carpenter et al., 2001, p. 766).

Although Carpenter et al. (2001) emphasise on the aspects of a socio-ecological system, they open the debate who or what makes the 'system' resilient. Therefore, the focus is shifted to regard the response of individuals in the state of shock in a context which is complex and shaped by various environments, highlighted by Adger (2000). Moreover, the definition of resilience by Carpenter et al. (2001) allows to perceive the response of individuals as a group which, it may also be referred to as community, is shaped by changes not only in the natural (ecological) environment but also in the political, economic and institutional dimensions, mentioned above.

Before discussing the resilience of communities in more detail a remark is needed to reveal that the concept of resilience is not solely used for ecological or socio-ecological approaches. Thus, in other fields of science, such as engineering, physics, psychology, etc., the concept of resilience is also adopted to describe, for example, the ability of a material to get back to its original shape after it got stretched. Manyena (2006) revisited the concept of resilience in detail and listed some of the numerous existing definitions of resilience (see Table 1) which was modified by the author.

Table 1. Definitions of Resilience.

Author	Definition
Wildavsky (1991)	Resilience is the *capacity* to *cope* with unanticipated dangers after they become manifest, learning to *bounce back*.
Holling, Schindler, Walker, and Roughgarden (1995)	It is the buffer *capacity* or the *ability* of a system to *absorb* perturbation, or the magnitude of disturbance that can be absorbed before a system changes its structure by changing the variables.
Horne and Orr (1998)	Resilience is a fundamental quality of individuals, groups and organisations and systems as a whole to *respond* productively to significant change that disrupts the expected pattern of events without engaging in an extended period of regressive behaviour.
Mallak (1998)	Resilience is the *ability* of an individual or organisation to expeditiously design and implement positive *adaptive* behaviours matched to the immediate situation, while enduring minimal *stress*.
Mileti (1999)	Local resiliency with regard to disasters means that a locale is able to *withstand* an extreme natural event without suffering devastating losses, damage, diminished productivity or quality of life without a large amount of assistance from outside the community.
Comfort et al. (1999)	The *capacity* to *adapt* existing resources and skills to new systems and operating conditions.
Paton, Smith, and Violanti (2000)	Resilience describes an active process of self-righting, learned resourcefulness and growth – the *ability to function* psychologically at a level far greater than expected given the individual's *capabilities* and previous experiences.
Kendra and Wachtendorf (2003)	The *ability to respond* to singular or unique events.
Cardona (2003)	The *capacity* of the damaged ecosystem or community to *absorb* negative impacts and recover from these.
Pelling (2003)	The *ability* of an actor to *cope* with or *adapt* to hazard *stress*.
RA (2007)	Three characteristics: firstly, the amount of change the system can undergo and still retain the same controls on function and structure; secondly, the degree to which the system is capable of self-organization and thirdly the *ability* to build and increase the *capacity* for learning and adaptation.
UNISDR (2009)	The *ability* of a system, community or society exposed to hazards to resist, *absorb*, accommodate to and recover from the effects of a hazard in a timely and efficient manner, including through the preservation and restoration of its essential basic structures and functions.

Source: Modified from Manyena (2006).

In the context of disasters, the concept of resilience has emerged in recent years to describe adaptation or the capacity to absorb an unexpected event. Therefore, based on the terminology from the United Nations International Strategy for Disaster Reduction (UNISDR), resilience is defined as 'the ability of a system, community or society exposed to hazards to resist, absorb, accommodate to and recover from the effects of a hazard in a timely and efficient manner, including through the preservation and restoration of its essential basic structures and functions' (UNISDR, 2009, p. 24).

As this definition is based on an amalgamation of available literature, it is not surprising that keywords of the concept of resilience, such as system, community, resist, absorb, recover, etc., are parts of this definition. As highlighted in Table 1, 'capacity', 'ability', 'withstand', etc. are some of the common terms widely used in the language describing aspects of resilience to disasters. Furthermore, in the context of disasters, abilities of adaptation and learning are widely regarded (Adger, Hughes, Folke, Carpenter, & Rockstrom, 2005; Folke, 2006; Klein, Nicholls, & Thomalla, 2003) as key elements characterising a resilient system.

Therefore, resilience is also associated to aspects of adaptive capacity (Cutter et al., 2008), which is described by various scholars as the ability of a system to learn from a disturbance by creating additional capacities to face a future disruption (Brooks, Adger, & Kelly, 2005; Burton, Saleemul, Lim, Pilifosova, & Schipper, 2002). Detailed explanations on concepts of vulnerability and adaptive capacity in the context of community resilience are given in the following sections and chapters.

Community Resilience to Disasters

The previous part emphasised on the wide use of resilience to describe processes where a high capacity to absorb a certain disturbance is regarded as resilient. In this section, the concept of resilience is brought down to the key objective of this dissertation which is, to understand the resilience of communities in urban areas to disasters.

Defining 'Community'

Before attempting to link the term community into the context of disaster and the concept of resilience, a differentiated understanding of what constitutes a community is needed. Chavis and Wandersman (1990) highlight two key types of 'communities': place-bound (with clear boundaries) and interest-based groups of people with no clear boundaries (Murphy, 2007).

However, both types are characterised as social systems where people have a certain degree of interdependence and interaction, developed networks and social capital. An example of a place-bound community in an urban area is a neighbourhood with defined political boundaries, like a ward, whereas an interest-based community is developed through the intense interaction of its members based on common interests (Murphy, 2007).

In the context of urban areas, Norris, Stevens, Pfefferbaum, Wyche, and Pfefferbaum (2008, p. 128) argue that 'communities are composed of built, natural, social, and economic environments that influence one another in complex ways'. Thus, communities may not be restricted to people-formed social units, but may also be understood in a wider sense, including 'non-human' aspects. As a result, the term community needs to be used carefully, as depending on the situation its meaning changes. For example, based on the scale of a city many different communities are included similar to a zone which is a cluster of a city (see Chapter 5). Furthermore, many different wards can refer to different communities. Thus, the community connotation is changing based on the scale: ward→zone→city.

Although this changing interpretation of 'community' may have the potential to cause some confusion, each of these types of communities refer equally to how physical, social, economic, institutional and natural characteristics perform within defined political boundaries.

Community Resilience

Attempting to connect the understanding of communities with the concept of resilience, various scholars (Adger, 2000; Allen, 2006; Bruneau et al., 2003; Paton & Johnston, 2001; Twigg, 2007) regard the extent of people's abilities to respond to a disturbance (e.g. disaster) to be shaped by the political, economic, physical and natural context of their environment where they are embedded in, as mentioned before. Thus, the initial thoughts originating from Holling's (1973) and Adger's (2000) work have been carried further amongst others by Bruneau et al. (2003) and Twigg (2007) who describe how the resilience of a community to a potential disaster is characterised. Accordingly, Twigg (2007, p. 6) defines it, as follows:

- 'capacity to absorb stress or destructive forces through resistance or adaptation;
- capacity to manage, or maintain certain basic functions and structures, during disastrous events;
- capacity to recover or "bounce back" after an event'.

Twigg's (2007) definition points out three phases for a community to be mastered successfully in order to be determined as fully resilient. For example, the capacity or ability of a community to absorb stress is largely relevant before a disaster occurs, it may also be defined as the coping capacity which includes the part where the system has to endure or manage the stress. Once, however, the disturbing event is over, the recovery or bouncing back phase begins. Hence, the term adaptive capacity defines to what extent the system in this case the community can get back to the pre-disaster situation or whether it fails to recover. In case the community can recover fully and go even beyond the pre-disaster resilience level, evidence of adaptive capacity is visible.

Bruneau et al. (2003), Cutter et al. (2008) and Tobin and Whiteford (2002) conceptualise community resilience as being not only relevant during an emergency, but put much more emphasis on the cyclic character of community resilience. That means a resilient community is one which is not only expected to be capable to absorb, maintain and recover from a shock but also adapt and increase its coping capacity to reduce the probability of being adversely affected by a future disaster (Holling & Gunderson, 2002; Pendall, Foster, & Cowell, 2010). This is highlighted by the Disaster Resilience of Place (DROP) model (Fig. 1) developed by Cutter et al. (2008) who emphasise on the cyclic character of community resilience. In other words, after a shock has occurred and a community, consisting of the built environment, natural and social systems, cannot fully cope with the event, a disaster may occur. Followed by the disaster shortly after, the community's

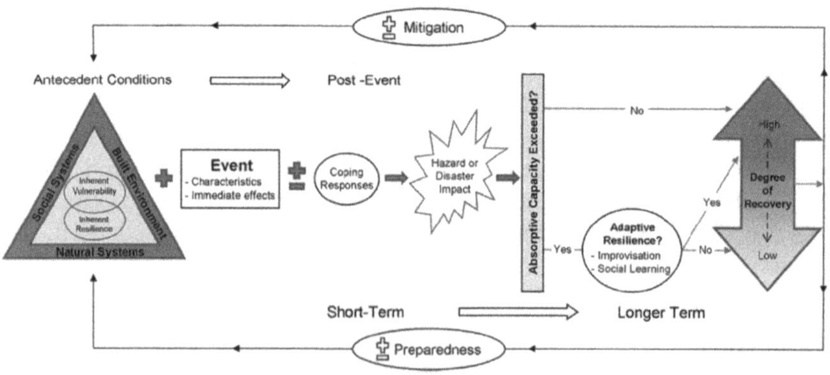

Fig. 1. Schematic Representation of the DROP Model. *Source*: Adapted from Cutter et al. (2008).

capacity to absorb and learn (adaptation) from the event determines to what extent it can recover. In the long term, the recovery process is replaced by the community's ability to mitigate and take measures for future disaster preparedness. The model itself is thus not static, but is much more dynamic as after a disaster is again before the next event. Consequently, the resilience of a community to disasters is a cyclic and has only reached the highest level once no natural hazards lead to disasters.

Interestingly, reflecting the term resilience from this perspective, it resembles the emergency/disaster management cycle with the four connected factors: preparedness, mitigation, response and recovery, described by King (2007). Therefore, resilience is a dynamic concept which is built and enhanced at any time, before, during and after a disaster. As Cutter et al. (2008) highlights the life-cycle character of a resilient community to disasters in the DROP model, the preparedness or non-disaster time gains particular importance.

For example, Bruneau et al. (2003) proclaim that a resilient community may have the potential to anticipate a disaster if it is adequately prepared and thus may even avoid a potential disaster.

To summarise so far, community resilience is dynamic and shaped by physical, social, economic, political and natural aspects characterising a defined area. The understanding of community resilience is further elaborated in Chapter 7.

Resilience versus Vulnerability

Before plunging into the discussion of how resilience in urban areas ought to be understood, it is necessary to differentiate 'resilience' from 'vulnerability' since improper distinction may lead to confusion (Cutter et al., 2008); although it may be difficult to clearly divide the two terms as they are to some extent related to each other (Berkes, Colding, & Folke, 2003). Looking at them from the time perspective, regarding to the advent of a disaster, a separation of vulnerability from resilience may be possible, whereby vulnerability is explained by Cutter et al. (2008, p. 599) as 'the pre-event, inherent characteristics or qualities of social systems that create the potential for harm'. Accordingly, the key parameters shaping the vulnerability of a system are its exposure, sensitivity and adaptive capacity to a potential harm (stress or shock) (Adger, 2006). Resilience, however, focuses more on a system's (community) response to an event (post-event), like whether it can resist, absorb or cope to it (Cutter et al., 2008). Furthermore, the 'ability to re-organise, change and learn in response to a threat' (Cutter et al., 2008, p. 599) shapes the level of a community's resilience.

Although an attempt is made to separate these two terms, the strong relationship between them must be emphasised. For example, the term adaptive capacity is used in both concepts (also discussed further in Chapter 7) and cannot be attributed to just one of the two concepts. Gallopin (2006), for example, describes adaptive capacity as the ability of socio-ecological systems, such as communities, to learn and improve their capacity either through reaction due to disturbance (e.g. disaster) or in a proactive manner where a future stress or change is anticipated before it occurs. While Gallopin's (2006) understanding may reflect aspects of resilience, Gunderson (2000) explains adaptive capacity through its ability to remain in a stable state. Thus, Gunderson's (2000) perception of adaptive capacity may rather replicate aspects of vulnerability where he puts little expectation in the system's ability to learn from the disturbance and become stronger to master a more intense future disturbance. At this point, the different understanding of the two concepts becomes apparent, in which resilience is regarded as a dynamic and interchanging.

However, the differences between vulnerability and resilience of socio-ecological systems (e.g. communities) are depending on the interpretation are often mixed and more research is needed to fully disentangle the two concepts. Although it is questionable whether this would be needed as at the end resilience and vulnerability aim to understand to what extent a system can manage with risk and disruptions.

APPLICATION OF THE CONCEPT OF RESILIENCE ON URBAN AREAS IN ASIA

In this section, two key issues are discussed: firstly, the exacerbating challenges imposed on cities due to impacts of climate change and urbanisation and secondly, the concept of resilience in relation to disasters and communities. In the following sections, the key objective is to link them together and provide the baseline for the Climate Disaster Resilience Index/Initiative (CDRI) assessment which is explained in more detail in Chapter 5.

The City as a Resilient System to Climate-Related Disasters

According to Godschalk (2003, p. 137), 'a resilient city is a sustainable network of physical systems and human communities' whereby both elements must be able to survive and function under extreme stresses (disasters). Similarly, Vale and Campanella (2005, p. 353) regard a disaster resilient city as 'a constructed phenomenon, not just in the literal sense that cities get reconstructed brick by brick, but in a broader sense'. Based on these initial definitions on how to understand a disaster resilient city, it becomes clear that a solid definition about what exactly determines a resilient city is difficult. This notion is given further ground when vague terms are used to describe the features of resilient city, which would enclose physical, social and institutional aspects (Klein et al., 2003; Norris et al., 2008).

The term resilient city is approached from the perspective to understand the strengths, weaknesses, abilities or capacities of communities (Bruneau et al., 2003; Cutter et al., 2008; Van Aalst, Cannon, & Burton, 2008), which are embedded in a densely built environment, to cope with climate-related disasters. The reason to approach a city's disaster resilience from a human-centric view is simply due to the circumstances that the overall objective is to understand the potential impacts of natural hazards. Thus, disasters are only felt if there is damage recorded on the built environment or human beings have been negatively affected by a natural hazard. Quarantelli (1985) provides crucial clarification on what constitutes a disaster and thereby emphasises on the different agents (physical, social, political, etc.) that may be harmed by a sudden disruption.

Going back to the conceptualisation of disaster resilience in urban areas, the perception that human beings are a key element of defining a community also recognises that the former is not seen in isolation to other disciplines and systems. Thus, the ability of human beings, living in a built environment (physical), to absorb and manage a disaster is shaped by political (institutional), economic and natural dimensions (Adger, 2000). The inherent character of city's with a high concentration of people also includes aspects of positive and negative aspects of social capital, marginalisation and social exclusion which are observed in communities.

In recent years, various new research teams/collaborations have linked the concept of resilience to cities. Largely due to the fact that many cities are growing rapidly. Hence, the Resilience Alliance (RA, 2007) defines resilience as the ability to absorb disturbances, to be changed and then to reorganise and still have the same identity (retain the same basic structures and ways of functioning). Thereby, resilience has three defining characteristics: (1) the amount of change the system can undergo and still retain the same controls on function and structure; (2) the degree to which the system is capable of self-organization and (3) the ability to build and increase the capacity for learning and adaptation. Based on these characteristics, the

Resilience Alliance started an urban resilience research programme that consists of four key elements: metabolic flow, social dynamics, governance network and built environment to better understand the resilience of cities to disasters (RA, 2007, 2009).

According to Surjan, Sharma, and Shaw (2011), there are several other recent studies on urban resilience; for example, the Megacity Resilience Framework (UNU-EHS, 2009) describes resilience as opposed to vulnerability due to its inability to cope with disaster risk. This framework further expands the definition as 'A (mega-) city can be regarded resilient if its inhabitants and institutions function effectively. That means that they are able to deal with unexpected disturbances and adapt to change. Furthermore, ecosystem services and their social and economic use by humans must be balanced. In this sense, the resilience of such a socio-ecological system is closely related to the concept of sustainability (economic, social and ecological)' (UNU-EHS, 2009, p. 3).

In another attempt, the Asian Cities Climate Change Resilience Network (ACCCRN, 2009) points to four elements of resilience: redundancy, flexibility, capacity to reorganize and capacity to learn. This network has been developed in 10 cities of 4 countries (Thailand, Vietnam, India and Indonesia) and has adopted a shared learning dialogue as a process-based approach at the city level. In continuing efforts to understand climate resilience planning in urban areas, climate resilient framework (Fig. 2) was developed to integrated aspects of understanding vulnerability, building resilience and knowledge which is linked and enhanced through shared learning. In this model, Moench, Tyler, and Lage (2011) aim to emphasise on the cyclic characteristics of building resilience where different systems (natural, economic, social, etc.) are interdependent. Again, the cyclic character of the urban climate resilience planning framework to build resilience is reaffirming Cutter et al.'s (2008) DROP model which assumes a dynamic conceptualisation of resilience rather than a static one-way understanding of how communities can increase their capacities to cope with disasters.

Finally, in a study of coastal community's resilience, United States Indian Ocean Tsunami Warning System Program (USIOTWSP, 2007) defines resilience through eight elements: governance, society and economy, coastal resource management, land-use and structural design, risk knowledge, warning and evacuation, emergency response and disaster recovery. This study argues that the community should be placed in the centre of the resilience concept, in order to understand how people adapt to changes through experience and apply lessons learned to enhance resilience.

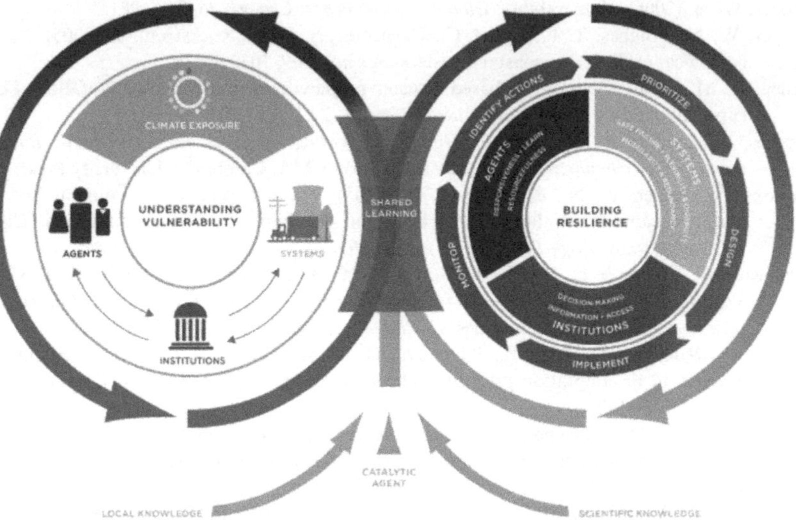

Fig. 2. Climate Resilient Framework. *Source*: Moench et al. (2011).

As all the above conceptualisations of resilience in urban areas highlighted, it is crucial to note that resilience is not a stable state but a dynamic concept with a cycle of its own which is once again exemplifying Cutter et al.'s (2008) DROP model. Hence, the resilience of an urban area is a complex function of different socio-economic, physical and institutional issues within a natural environment. The above literature review also points out that resilience of an urban community depends on its strength to form the ideal environment, which is best capable to minimise the probability of shocks and has the highest ability to respond to situations of disaster. Finally, based on this conceptualisation of resilience to disasters in urban areas, the CDRI, presented in Chapter 5, takes up these issues (physical, social, economic, institutional and natural) that are determining how well a community can cope and adapt to disasters.

REFERENCES

ACCCRN. (2009). *Asian Cities Climate Change Resilience Network (ACCCRN): Responding to the urban climate challenge*. Boulder, CO: ISET.

Adger, W. N. (2000). Social and ecological resilience: Are they related? *Progress in Human Geography*, *24*(3), 347–364.

Adger, W. N. (2006). Vulnerability. *Global Environmental Change, 16*, 268–281.

Adger, W. N., Hughes, T. P., Folke, C., Carpenter, S. R., & Rockstrom, J. (2005). Social-ecological resilience to coastal disasters. *Science, 309*, 1036–1039.

Allen, K. M. (2006). Community-based disaster preparedness and climate adaptation: Local capacity-building in the Philippines. *Disasters, 30*(1), 81–101.

Berkes, F., Colding, J., & Folke, C. (2003). *Navigating social–ecological systems: Building resilience for complexity and change*. Cambridge, MA: Cambridge University Press.

Brooks, N., Adger, W. N., & Kelly, M. P. (2005). The determinants of vulnerability and adaptive capacity at the national level and the implications for adaptation. *Global Environmental Change Part A, 15*(2), 151–163.

Bruneau, M., Chang, S. E., Eguchi, R. T., Lee, G. C., O'Rourke, T. D., Reinhorn, A. M., ... Von Winterfeldt, D. (2003). A framework to quantitatively assess and enhance the seismic resilience of communities. *Earthquake Spectra, 19*(4), 733–752.

Burton, I., Saleemul, H., Lim, B., Pilifosova, O., & Schipper, E. L. (2002). From impacts assessment to adaptation priorities: The shaping of adaptation policy. *Climate Policy, 2*(2–3), 145–159.

Cardona, O. D. (2003). *The notions of disaster risk: Conceptual framework for integrated management. information and indicators program for disaster risk management.* Manizales: Inter-American Development Bank.

Carpenter, S., Walker, B., Anderies, J. M., & Abel, N. (2001). From metaphor to measurement: Resilience of what to what? *Ecosystems, 4*, 765–781.

Chavis, D. M., & Wandersman, A. (1990). Sense of community in the urban environment: A catalyst for participation and community development. *American Journal of Community Psychology, 18*(1), 55–81.

Comfort, L., Wisner, B., Cutter, S., Pulwarty, R., Hewitt, K., Oliver-Smith, A., ... Krimgold, F. (1999). Reframing disaster policy: The global evolution of vulnerable communities. *Environmental Hazards, 1*, 39–44.

Cutter, S. L., Barnes, L., Berry, M., Burton, C., Evans, E., Tate, E., & Webb, J. (2008). A place based model for understanding community resilience to natural disasters. *Global Environmental Change, 18*, 598–606.

Folke, C. (2006). Resilience: The emergence of a perspective for social-ecological system analyses. *Global Environmental Change, 16*, 253–267.

Gallopin, G. C. (2006). Linkages between vulnerability, resilience, and adaptive capacity. *Global Environmental Change, 16*, 293–303.

Godschalk, D. R. (2003). Urban hazard mitigation: Creating resilient cities. *Natural Hazards Review, 4*(3), 136–143.

Gunderson, L. (2000). Ecological resilience – In theory and application. *Annual Review of Ecological Systems, 31*, 425–439.

Holling, C. S. (1973). Resilience and stability of ecological systems. *Annual Review of Ecological Systems, 4*, 1–23.

Holling, C. S., & Gunderson, L. H. (2002). Resilience and adaptive cycles. In L. H. Gunderson & C. S. Holling (Eds.), *Panarchy: Understanding transformations in human and natural systems* (pp. 27–33). Washington, DC: Island Press.

Holling, C. S., Schindler, D. W., Walker, B. W., & Roughgarden, J. (1995). Biodiversity in the functioning of ecosystems: An ecological synthesis. In C. Perrings, K. G. Maler, C. Folke, C. S. Holling, & B. O. Jansson (Eds.), *Biodiversity loss: Economic and ecological issues* (pp. 44–83). Cambridge, MA: Cambridge University Press.

Horne, J. F., & Orr, J. E. (1998). Assessing behaviours that create resilient organisations. *Employment Relations Today*, *24*(4), 29–39.

Kendra, M. J., & Wachtendorf, T. (2003). Elements of resilience after the world trade center disaster: Reconstructing New York city's emergency operation center. *Disasters*, *27*(1), 37–53.

King, D. (2007). Organisations in disaster. *Natural Hazards*, *40*, 657–665.

Klein, R. J. T., Nicholls, R. J., & Thomalla, F. (2003). Resilience to natural hazards: How useful is this concept? *Environmental Hazards*, *5*, 35–45.

Mallak, L. (1998). Resilience in the healthcare industry. Paper presented at the Seventh Annual Engineering Research Conference, 9–10 May, Banff.

Manyena, S. B. (2006). The concept of resilience revisited. *Disasters*, *30*, 433–450.

Mileti, D. S. (1999). *Disasters by design: A reassessment of natural hazards in the United States*. Washington, DC: Joseph Henry Press.

Moench, M., Tyler, S., & Lage, J. (2011). *Catalyzing urban climate resilience. Applying resilience concepts to planning practice in the ACCCRN program (2009–2011)*. Boulder, CO: Institute of Social and Environmental Transition (ISET), International ISET.

Murphy, B. L. (2007). Locating social capital in resilient community-level emergency management. *Natural Hazards*, *41*, 297–315.

Norris, F. H., Stevens, S. P., Pfefferbaum, B., Wyche, K. F., & Pfefferbaum, R. L. (2008). Community resilience as a metaphor, theory, set of capacities, and strategy for disaster readiness. *American Journal of Community Psychology*, *41*, 127–150.

Paton, D., & Johnston, D. (2001). Disasters and communities: Vulnerability, resilience and preparedness. *Disaster Prevention and Management*, *4*, 270–277.

Paton, D., Smith, L., & Violanti, J. (2000). Disasters response: Risk, vulnerabilities and resilience. *Disaster Prevention and Management*, *9*(3), 173–179.

Pelling, M. (2003). *The vulnerability of cities: Natural disasters and social resilience*. London: Earthscan.

Pendall, R., Foster, K. A., & Cowell, M. (2010). Resilience and regions: Building understanding of metaphor. *Cambridge Journal of Regions, Economy and Society*, *3*, 71–84.

Quarantelli, E. L. (1985). What is disaster? The need for clarification in definition and conceptualization in research. In B. Sowder (Ed.), *Disasters and mental health selected contemporary perspectives* (pp. 41–73). Washington, DC: Government Printing Office.

Resilience Alliance (RA). (2007). *A research prospectus for urban resilience. A resilience alliance initiative for transitioning urban systems towards sustainable futures*. Retrieved from http://www.resalliance.org/1610.php. Accessed on June 13, 2010.

Resilience Alliance (RA). (2009). *The megacity resilience framework*. Retrieved from http://www.ehs.unu.edu/file/get/4074. Accessed on November 30, 2010.

Surjan, A., Sharma, A., & Shaw, R. (2011). Understanding urban resilience. In R. Shaw & A. Sharma (Eds.), *Climate and disaster resilience in cities* (pp. 17–45). Bingley: Emerald.

Tobin, G. A., & Whiteford, L. A. (2002). Community resilience and volcano hazard: The eruption of Tungurahua and evacuation of the faldas in Ecuador. *Disasters*, *26*(1), 28–48.

Twigg, J. (2007). *Characteristics of a disaster-resilient community: A guidance note. Disaster Risk Reduction Interagency Coordination Group*. DFID: Benfield.

United Nations International Strategy for Disaster Reduction (UNISDR). (2009). *Global assessment report on disaster risk reduction*. Geneva: United Nations.

United States Indian Ocean Tsunami Warning System Program (USIOTWSP). (2007). *How resilient is your coastal community? A guide for evaluating coastal community resilience to tsunamis and other coastal hazards*. Bangkok: USIOTWSP.

UNU-EHS. (2009). *Policy brief 'The Megacity Resilience Framework'*. Retrieved from http://www.ehs.unu.edu/article/read/policy-brief. Accessed on October 12, 2010.

Vale, L., & Campanella, T. (2005). *The resilient city: How modern cities recover from disaster*. New York, NY: Oxford University Press.

Van Aalst, M. K., Cannon, T., & Burton, I. (2008). Community level adaptation to climate change: The potential role of participatory community risk assessment. *Global Environmental Change*, *18*, 165–179.

Wildavsky, A. (1991). *Searching for safety*. New Brunswick: Transaction.

CHAPTER 4

RESILIENCE IN THE CONTEXT OF URBAN DISASTER RISK REDUCTION IN INDIA

INTRODUCTION

In this chapter, the term resilience is put into the context of Disaster Risk Reduction (DRR). First of all, it must be noted that the understanding of DRR is relatively new and evolved only over the past two to three decades. Nonetheless, the current definition provided by the United Nations International Strategy for Disaster Reduction (UNISDR) describes key issues of DRR, as follows: 'the concept and practice of reducing disaster risks through systematic efforts to analyse and manage the causal factors of disasters, including through reduced exposure to hazards, lessened vulnerability of people and property, wise management of land and the environment, and improved preparedness for adverse events' (UNISDR, 2009, pp. 10–11). As this chapter will later explain, this DRR definition is an amalgamation of key elements associated to how disasters ought to be addressed according to the adopted Hyogo Framework for Action (HFA) in 2005.

Following the description of the current status of international efforts on the implementation of DRR, this chapter explains how DRR is currently addressed and implemented in urban areas of India. Thereby, attention is given to how DRR is facilitated and mainstreamed at the local level in India. Finally, the chapter aims to draw a linkage between DRR and resilience where the former is expected to enhance the capacity of an urban area to absorb or lessen the probability of climate-related disasters.

INTERNATIONAL EFFORTS ON DISASTER RISK REDUCTION

Origin of Disaster Risk Reduction

During the 1980s, the US Academy of Sciences brought up the idea to launch an international decade addressing the impacts from growing numbers of devastating disasters (Verstappen, 1993). This led to the adoption of an UN resolution in April 1989 to form the International Decade for Natural Disaster Reduction (IDNDR) (United Nations Centre for Regional Development (UNCRD), 1989). The key aspirations declared in Tokyo Declaration are as follows:

> The people of the world, as well as their governments, to work toward greater security against natural disaster;
>
> The governments of all countries to participate actively in the Decade by educating and training their citizens to increase awareness, by enhancing social preparedness, by integrating disaster-consciousness into their development programmes, and by making available the power of science and technology to reduce disaster loss;
>
> The United Nations, scientific and technological institutions, non-governmental organizations, and the private sector to support international and regional cooperation on disaster-related activities and to contribute to the transfer of disaster-reduction technology, particularly in disaster-prone developing countries. (UNCRD, 1989, p. 1)

In particular, the second bullet points out the need for reducing disaster losses through more integrative (including social issues) planning of development. In 1992, the UN World Summit in Rio de Janeiro raised another crucial issue related to DRR which is about more cautious usage of depleting natural resources, in response to the Brundtland report (1987), the declaration of this conference addressed the need for sustainable development and called for environmental protection to be an integral part in it (United Nations (UN), 1992). Accordingly, the burning of fossil fuels was seen as detrimental to the climate and thus the term climate change was essentially born and recognised by the participating nations. Concrete outcomes of this Earth Summit were a climate change convention that addressed the need to take action on minimising the potential consequences of climate change through efforts focusing on adaptation and mitigation. Also a convention on biological diversity was adopted to develop ecoregions to support the protection of habitats of various species that may become endangered due to unsustainable development.

Finally, apart from convention on climate change and biodiversity, Principle 18 of the conference declaration stated the following issue in relation to disasters: 'States shall immediately notify other States of any natural disasters or other emergencies that are likely to produce sudden harmful effects on the environment of those States. Every effort shall be made by the international community to help States so afflicted' (UN, 1992). Thus, the focus in this principle was more on providing sufficient response rather than taking preventive or pro-active measures to avoid disasters.

Yokohama Strategy

Two years later in 1994 at the World Conference on Natural Disaster Reduction, the Yokohama Strategy was unveiled and reaffirmed the Principle 18 from the UN World Summit in 1992, but much more founded a new approach to regard disasters caused by natural hazards in a more dynamic way. While losses on human lives and infrastructures has risen over time, as mentioned already in the Brundtland report and addressed in the Rio declaration, a disaster was looked from the perspective that prevention, preparedness and mitigation measures could avoid or lessen the impacts from such an event (IDNDR, 1994). Essentially, the idea at that time was that disasters (Yokohama Strategy) as well as climate change (Kyoto Protocol, 1998) could be avoided if the right steps would be taken. Although the Yokohama Strategy emphasised on building enhanced capacities among communities, technical solutions to lessen the probability of disasters were largely encouraged. Accordingly, the Kyoto Protocol focused on reducing greenhouse gas emissions. Nonetheless, the Yokohama Strategy laid the foundation to perceive disasters from a broader point of view where the causes of disasters needed to be measured (risk assessment) and proper action taken to reduce disaster risk.

Ten years later in the preparation to the next World Conference, the Yokohama Strategy was reviewed by an UN-led Inter Agency Task Force on Disaster Reduction in consultation with other national governments and non-governmental organisations (NGOs) during various sessions and presented the following key challenges that so far did not get sufficient attention (UN, 2004):

(a) Governance: organisational, legal and policy frameworks;
(b) Risk identification, assessment, monitoring and early warning;
(c) Knowledge management and education;
(d) Reducing underlying risk factors;
(e) Preparedness for effective response and recovery.

The gaps identified above essentially urged to develop disaster reduction strategies that are tailored to each countries institutional, economic and social structures. In other words, the mainstreaming of DRR into all types of plans and policies at the national level were advocated as primary steps to reduce the risks from disasters (UN, 2004). As the latter sentence reflected point a) from above, each point was addressed in detail and formed a new non-binding policy named HFA.

Hyogo Framework for Action 2005–2015

Almost 11 years after the Yokohama Strategy, the HFA 2005–2015 – Building the Resilience of Nations and Communities to Disasters – was adopted by 168 nations during the World Conference on Disaster Reduction in Kobe in 2005. Based on the review from the Yokohama Strategy, the HFA incorporated the key challenges mentioned above and declared them as principles that are framing the new strategy on how best disasters can be avoided or its impacts lessened (UNISDR, 2005).

Following the adoption of the HFA, a detailed guidance on how to implement this policy at the national level was presented in the document Words Into Action in 2007 (UNISDR, 2007). These five key principles were supported by an additional 22 tasks outlining the content of the principles. As the HFA's primary intention is to mainstream national platforms on disaster management, little emphasis was initially given to how action should be taken at the local level. However, through minor modifications of the initial HFA, a guide was developed defining the actions that ought to be taken at the local level (Kyoto University (KU), 2010). Table 1 highlights the five principles and 20 tasks that are directed towards city governments. The need for this localised document is basically due to the fact that action to prevent and manage a potential disaster is usually undertaken at the local level. In other words, the provision of relief and emergency response during a disaster is usually handled by local people. Thus, the localised HFA (KU, 2010) supports actions at this level of governance.

The transformation of the original (national focus) HFA to the local HFA further supports a paradigm shift that participatory bottom-up efforts are regarded as crucial to implement DRR. Furthermore, the identified tasks in the local HFA provide an important orientation in the process of developing new risk assessment tools, such as the CDRI presented in Chapter 5. In particular, HFA Priority 4 emphasises on the need to address so-called underlying risk which includes a large array of measures ranging from environmental, social, physical and economic issues that are prevalent in cities and villages (see Table 1).

Table 1. Twenty Tasks of Local HFA.

	Local/City Governance (HFA Priority 1 Related)
Task 1	Engage in multi-stakeholder dialogue to establish foundations for DRR
Task 2	Create or strengthen mechanisms for systematic coordination for DRR
Task 3	Assess and develop the institutional basis for DRR
Task 4	Prioritize DRR and allocate appropriate resources
	Risk Assessment and Early Warning (HFA Priority 2 Related)
Task 5	Establish an initiative for community risk assessment to combine with country assessments
Task 6	Review the availability of risk-related information and the capacities for data collection and use
Task 7	Assess capacities and strengthen early warning systems
Task 8	Develop communication and dissemination mechanisms for disaster risk information and early warning
	Knowledge Management (HFA Priority 3 Related)
Task 9	Raise awareness of DRR and develop education programme on DRR in schools and local communities
Task 10	Develop or utilize DRR training for key sectors based on identified priorities
Task 11	Enhance the compilation, dissemination and use of DRR information
	Vulnerability Reduction (HFA Priority 4 Related)
Task 12	Environment: Incorporate DRR in environmental management
Task 13	Social needs: Establish mechanisms for increasing resilience of the poor and the most vulnerable
Task 14	Physical planning: Establish measures to incorporate DRR in urban and land-use planning
Task 15	Structure: Strengthen mechanisms for improved building safety and protection of critical facilities
Task 16	Economic development: Stimulate DRR activities in production and service sectors
Task 17	Financial/economic instruments: Create opportunities for private sector involvement in DRR
Task 18	Emergency and public safety; disaster recovery: Develop a recovery planning process that incorporates DRR
	Disaster Preparedness (HFA Priority 5 Related)
Task 19	Review disaster preparedness capacities and mechanisms, and develop a common understanding
Task 20	Strengthen planning and programming for disaster preparedness

Source: KU (2010).

Progress of HFA Implementation

After the adoption of the HFA in 2005, the leading UN agency, the UNISDR, undertook evaluation procedures to measure the implementation level of the five HFA priority areas in all the countries which are taking part in the HFA. After the first Global Assessment Report (GAR) was published in 2009, a more comprehensive report focusing, in particular, on the implementation progress of the HFA at the national level of countries was presented in 2011. The report highlights that HFA Priorities 3 and 4, concerning about issues related to knowledge and education and underlying risk are progressing slower than the other priorities. Thus, more efforts are needed regarding raising awareness among people and alleviate disaster risk.

The CDRI assessments described in Chapter 5 have the objective to address the HFA priorities, in particular, the various underlying risk. The 2011 GAR on DRR makes clear that if the climate disaster risk, discussed in Chapter 2, are not given more emphasis in planning processes, they may lead to intensive risks and, therefore, become key elements triggering disasters (UNISDR, 2011). Critical countries with slower progress include largely developing countries in Latin America, Africa and Asia (UNISDR, 2011).

Despite the fact that only 82 out of 133 participating governments shared their interim reports of the HFA implementation (period from 2005 to 2015) in the 2011 GAR, the efforts undertaken at the international level support the implementation of DRR.

United Nations' 2010–2011 World Disaster Reduction Campaign

The previous section 'Origin of Disaster Risk Reduction' emphasised on the origins of DRR and efforts undertaken by the international community which culminated in the adoption of the HFA in 2005. To sustain the ongoing and to stimulate further activities regarding DRR in urban areas, the 2010–2011 World Disaster Reduction Campaign – Making Cities Resilient – provided a platform for cities to undertake self-assessments about their cities' resilience to disasters.

Table 2 highlights the 10-point checklist which reflects key aspects of the five HFA priorities. As of October 2011, 904 cities all around the world participated in this campaign, including Chennai. More details on the scores of the 10-point checklist and progress of DRR in Chennai are discussed in Chapter 8. However, a major challenge especially for developing countries is the assigning of budget for DRR (checkpoint 2), according to the GAR (UNISDR, 2011). Thus, low-income countries – not surprisingly – find it difficult to provide specific funding for reducing disaster risk due to other priorities. This fact calls for a paradigm shift to re-orientate DRR funding away from specific budget allocations. Instead DRR should

Table 2. Ten-Point Checklist: Making Cities Resilient.

1. Put in place *organization and coordination* to understand and reduce disaster risk, based on participation of citizen groups and civil society. Build local alliances. Ensure that all departments understand their role to DRR and preparedness.
2. *Assign a budget* for DRR and provide incentives for homeowners, low-income families, communities, businesses and public sector to invest in reducing the risks they face.
3. Maintain up-to-date data on hazards and vulnerabilities, *prepare risk assessments* and use these as the basis for urban development plans and decisions. Ensure that this information and the plans for your city's resilience are readily available to the public and fully discussed with them.
4. Invest in and maintain *critical infrastructure that reduces risk*, such as flood drainage, adjusted where needed to cope with climate change.
5. Assess the *safety of all schools and health facilities* and upgrade these as necessary.
6. Apply and enforce *realistic, risk-compliant building regulations and land use planning principles*. Identify safe land for low-income citizens and develop upgrading of informal settlements, wherever feasible.
7. Ensure *education programmes and training* on DRR are in place in schools and local communities.
8. *Protect ecosystems and natural buffers* to mitigate floods, storm surges and other hazards to which your city may be vulnerable. Adapt to climate change by building on good risk reduction practices.
9. Install *early warning systems and emergency management capacities* in your city and hold regular public preparedness drills.
10. After any disaster, ensure that the *needs of the survivors are placed at the centre of reconstruction* with support for them and their community organizations to design and help implement responses, including rebuilding homes and livelihoods.

Source: UNISDR (2010).

be made an integrative part of all new developments as well as in retrofitting activities of existing infrastructures. This would, however, require that the private sector and communities become more strongly involved into DRR implementation processes.

Apart from this World Disaster Reduction Campaign focusing on DRR implementation in urban areas, various transnational city organisations/associations strengthen their efforts to make DRR implementation a top priority. Examples of such bodies are the Arab Town Organisation, CITYNET or ICLEI. These organisations are supporting the sharing of knowledge on how DRR is actually implemented in cities.

DRR in Other International and Regional Organisations

Apart from the UN efforts on making resilience a priority in urban areas through the above-mentioned World Disaster Reduction Campaign, other international and regional (focus in Asia) organisations advocate the inclusion of DRR into urban development plans, policies and programmes.

One example is the Global Facility on Disaster Risk Reduction (GFDRR) which is managed by the World Bank and deals with issues related to disaster management. Accordingly, this organisation has set up three tracks:

- Track I: Global and Regional Partnerships;
- Track II: Mainstreaming DRR;
- Track III: Sustainable Recovery.

Among the three pathways, DRR is particularly advocate in the second track which follows the five priorities from the HFA. The GFDRR is recently increasing its operations across the world to support the mainstreaming (planning, implementation) of DRR not only at the national level of countries but also at the local level through partnerships.

Regarding regional organisations, the Association of Southeast Asian Nations (ASEAN) promotes and advocates for safer communities. In their current Roadmap for ASEAN Community (2009–2015), the objective is to create disaster-resilient communities through the promotion of livelihood options which have the potential to increase socio-economic activities that are minimising disaster risk and strengthen community capacities (ASEAN, 2009). Furthermore, DRR is recognised as a component that requires integration into development plans at all levels (including urban areas).

Another crucial regional multi-national organisation is the South Asian Association for Regional Cooperation (SAARC) which includes eight countries in South Asia. Despite little concrete efforts on DRR, this organisation is aiming to tackle issues related to DRR, such as poverty, gender equality, etc., in a collaborative effort between its member states (SAARC, 2007). The current SAARC Development Goals have the primary focus to enhance collaboration between the SAARC members; however, the potential of SAARC as a regional body, integrating some of the poorest and heavy disaster-prone countries of the world, which can advocate, and mainstream DRR in urban areas, requires to be considered.

DRR IN INDIA – NATIONAL LEVEL

Before going into more detail about discussing DRR in India, it is important to note that as of today there is no explicit national strategy on DRR available in India. However, that does not mean that DRR is inexistent in India, as the following sections highlight.

Disaster Management Act

In 2005, the national legislative body of India enacted the country's first Disaster Management Act (DMA). The key features of this act are as follows: firstly, the country's institutional setup related to disasters is reformed and requires now each state and district to develop a body that is responsible to manage issues concerning disasters (Government of India (GoI), 2005). Secondly, more power is dissolved from the national level to the states and districts which are now obliged to develop disaster management plans for their constituencies. Finally, efforts enhancing the preparedness and capacity of communities are recognised as viable measures to increase the country's action for disaster preparedness and mitigation.

Although necessary legislative changes have been introduced through the implementation of the DMA and more power is given to the state and local level authorities, the country still lacks the proper implementation of this act. This means that many states still do not have developed disaster management plans and in some states neither disaster management authorities at the state level exist. Such an example is the State of Tamil Nadu which as of today still has no functioning disaster management authority and thus no disaster management plan. Despite its formal establishment in 2008 (Government of Tamil Nadu, 2008), the authority has of today not yet been functioning effectively.

In contrast, the State of Gujarat already established a disaster management authority, Gujarat Disaster Management Authority (GSDMA), in 2001 long before the implementation of the national DMA in 2005. Although a disaster management plan is still not available at the state level in Gujarat, the authority developed the Gujarat State Disaster Management Policy (GSDMP) in 2002, shortly after the establishment of the authority, which served to the formulation of India's first state to have a DMA implemented in 2003 (GSDMA, 2003). In this DMA, the need to reduce disaster risk is recognised; however, little emphasis is put on issues related to adaptation. Furthermore, DRR is also not yet recognised as a mechanism to be mainstreamed in all development plans and policies.

Another comparatively 'active' state is Himachal Pradesh in the north of India which has presented its Disaster Management Policy (DMP) in 2011 (Government of Himachal Pradesh (GHP), 2011a) and presented a state-level disaster management plan in 2011 (GHP, 2011b) in collaboration with SEEDS India (a nationally recognised NGO engaged in the field of DRR). Unlike the DMP from the State of Gujarat, the equivalent policy in Himachal Pradesh recognises DRR and advocates for its mainstreaming in all plans, policies and programmes.

Finally, the State of Maharashtra must be regarded as another 'active' state with a functioning disaster management authority. Furthermore, in September 2011, a comprehensive state disaster management action plan has been published which focuses on various aspects related to DRR, such as risk and vulnerability assessments, and also recognises the involvement and participation of communities and the civil society (e.g. NGOs) as key partners for an effective management of disasters.

To summarise, the need for national legislation on disaster management has been addressed in the implementation of a DMA in 2005, but so far few states have followed the expectations formulated in this act. Therefore, not surprisingly, DRR is also given little emphasis at all levels of governance in India. Nevertheless, DRR is getting recognised, for example, in Himachal Pradesh as a mechanism that requires becoming part of all plans, policies and programmes.

Organisational Setup for Potential DRR Implementation

The previous section highlighted that as of today comparatively little concrete legislative efforts have been done to foster DRR. Thus, the availability and existence of DRR in India should be approached from a different perspective.

Looking at Table 3, organisations with a potential involvement in Urban Disaster Risk Reduction (UDRR) are listed; thereby, the focus is unavoidably put on organisations acting at the national level. Due to the hierarchical institutional system in India, the planning for UDRR is concentrated in country-level initiatives which will be discussed in the following sections.

Table 3 emphasises on the importance of the Ministry of Urban Development as a key institutional body to include DRR in its policies related to the development of India's urban areas. Accordingly, one of their key project or mission is the Jawaharlal Nehru Urban Renewal Mission (JNNURM) which was launched in 2005.

Apart from organisations functioning under the supervision of different ministries, there are also semi-independent authorities such as the National Disaster Management Authority (NDMA) and National Institute of

Table 3. Organisations Involved in UDRR at Various Institutional Levels in India.

Level	Urban Planning/Development	Disaster Management
National level	a) Ministry of Urban Development • Central Public Works Department • Town and Country Planning Organisation • National Institute of Urban Affairs Schemes under JNNURM: – Urban Infrastructure and Governance (UIG) – Urban Infrastructure Development Scheme for Small and Medium Towns – North Eastern Region Urban Development Programme (NERUDP) b) Ministry of Housing and Urban Poverty Alleviation Scheme under JNNURM: – Basic Services to the Urban Poor (BSUP)	a) Ministry of Home Affairs • National Disaster Management Division b) National Disaster Management Authority • National Institute of Disaster Management
State level (key departments)	a) Housing and Urban Development Department b) Municipal Administration and Water Supply Department (Tamil Nadu) c) Public Works (Tamil Nadu) d) Highways and Minor Ports (Tamil Nadu)	Department of Revenue: • Special Relief Commissioner
District/city level	Town planning unit offices in some districts: a) Development authority (in few cities) b) Municipal Corporation	a) District Disaster Management Authority b) Public Works Department (partially) c) Fire Department (partially)

Source: Modified from Surjan (2008).

Disaster Management (NIDM) which are key organisations in the process of DRR implementation. In particular, the NDMA is currently involved in the execution of a national programme introduced by the GoI and the United Nations Development Programme (UNDP) to foster DRR in during the period from 2009 to 2012 through various projects in all the 28 states and Delhi (one out of seven union territories) of India (NDMA, no dateA).

The aim of this programme is to support and strengthen central and state governments' institutional structures to stimulate effective DRR (NDMA, no dateB). Furthermore, the programme supports projects that focus on reducing disaster risk and enhancing preparedness and recovery capacities. The total budget for this programme is USD 20 million and is expected to get further insights through various pilot studies carried out in all parts of India. According to the Progress Report (NDMA, no dateB), most activities financed through this programme focused on the training of various stakeholders (governmental staff, NGOs and communities).

Although this programme does not focus specifically on urban areas, it allows understanding the institutional challenges and nuisances related to DRR implementation in India.

India's Five-Year Development Plans

As effective DRR implementation requires its mainstreaming into all governmental plans, policies and programmes, the five-year plan from the GoI gains particular importance in this process, as it constitutes the key plan determining India's development. After India's independence from the British Empire in 1947, the first five-year plan was introduced in 1951 to lay out the development strategies of the central government.

Since DRR is a rather new concept, it was mentioned for the first time in the 10th Five-year Plan (2002–2007) where it was emphasised to be mainstreamed into development processes (GoI, 2008). In the following 11th Five-year Plan (2007–2012), DRR is further encouraged to be adopted by corporate sectors, NGOs and individuals (GoI, 2008). Thus, for example, government led incentives to provide tax cuts for individuals who retrofit unsafe buildings. Furthermore, initiatives strengthening Public Private Partnerships (PPP) are regarded as viable approaches for effective DRR.

However, despite the government's clear statement to integrate DRR into development process across different sectors, the actual implementation relies upon the state and local governments. This process has been activated, but further efforts are required to, particularly in raising awareness among individuals to let DRR break through. The CDRI approach and subsequent studies in Chapters 5–7 aim to address these challenges related to the context of Indian cities in general and, in particular, in Chennai.

Jawaharlal Nehru National Urban Renewal Mission

A key programme fostering urban renewal projects and a potential provider of DRR is the JNNURM (see schemes under JNNURM in Table 4)

Table 4. Schemes under the JNNURM.

Scheme	Year	Funding Source	Key Objective
Urban Infrastructure and Governance (UIG)	From 2005–	100% Central Government → loan and grant	• Improvement of urban services (water, sanitation, roads, etc.) in cities covered under the Mission (*only 63 selected cities*) • Tackling outgrowth into pero-urban areas/ urban corridors due to dispersed urbanisation • Urban renewal programme for the old city areas to reduce congestion
UIDSSMT	From 2005 to 2012	80–90% Central Government 10–20% State Government → grant	Small and medium cities, not covered under UIG and BSUP: • Improvement of urban services and help the creation of durable public assets and quality oriented services in cities & towns • Enhancement of public-private-partnership in infrastructural development and • Promotion of planned integrated development of towns and cities
North Eastern Region Urban Development Programme (NERUDSP)	From 2009 to 2015	Central Government and Asian Development Bank → loan: USD 274 million (divided into three tranches)	Capital cities of 5 North Eastern states: • Improvement of urban services: water supply, sewerage and sanitation and solid waste management • Project management and capacity development of ULBs
BSUP	From 2005–	See Table 6	Covers only selected cities (63): • Provision of basic services to the urban poor including security of tenure at affordable prices, improved housing, water supply and sanitation

Source: GoIMHUPA (2009), GoIMUD (2009), JNNURM (2009), NERUDP (no date).

established in 2005 led by the Ministries of Urban Development and Housing and Urban Poverty Alleviation (see Table 3). As Chapter 2 emphasised, India's urban areas are challenged by exhausted basic services; therefore, the JNNURM has the viable objectives to provide relief from impacts of urbanisation through the financing of projects that aim to improve the road networks, public transport, drainage and sanitation systems, water supply, etc. (Government of India Ministry of Urban Development (GoIMUD), 2011a; JNNURM, 2009).

Apart from financing urban infrastructure projects, the JNNURM further has the objective to strengthen community participation and the accountability of ULBs towards their citizens (JNNURM, 2009). The duration of this mission was initially set until 2012–2013, but is likely to extend and will become an integral part of the 12th Five-year Plan which is currently under consultation. As of September 2011, 536 projects have been approved at an estimated cost of USD 12.28 billion (JNNURM, 2011). Thus, only around half of the estimated investment requirements have been approved so far (see Table 5).

Urban Infrastructure and Governance

Although the JNNURM aims to improve India's urban areas, Urban Infrastructure and Governance (UIG) scheme only includes city with high populations above 1 million and a selected number of cities with less than 1 million (see Table 5). As mentioned earlier, the draft 12th Five-year Plan

Table 5. Indian Urban Sector Investment Requirement under UIG Scheme.

Category	Number of Cities	Investment Requirement (over 7 Years Starting 2005–2006; in billion USD)	Annual Funds Requirement (in billion USD)
Cities with over 4 million population	7	11.56[a]	1.65[a]
Cities with 1–4 million population	28	11.56[a]	1.65[a]
Selected cities with less than 1 million population	28	1.26[a]	0.18[a]
Total	**63**	**24.38**	**3.48**

Source: JNNURM (2009).

[a]Costs converted from Indian Rupees into USD at an exchange rate of 1 Rupee = 0.02 USD on 4 November 2011.

addresses the need for enhanced PPPs and stronger ULBs (GoI, 2011). Furthermore, it projects that the costs for improvements of urban infrastructures to just absorb the impacts of urbanisation would be as high as USD 79.3 billion over the next 20 years (GoI, 2011). The magnitude of this cost estimation makes visible the sheer amount of investment required in the future to maintain a certain quality of life in India's urban areas.

Urban Infrastructure Development Scheme for Small and Medium Towns
Another important scheme for urban areas and cities is the Urban Infrastructure Development Scheme for Small and Medium Towns (UIDSSMT) which is rather similar to the UIG and has almost the same objectives; it was also launched in 2005 by the Ministry of Urban Development under the JNNURM mission (GoIMUD, 2011b). However, it excludes all the 63 cities under the UIG scheme. Its objectives are as follows: firstly, improve infrastructural facilities and help create durable public assets and quality-oriented services in cities and towns; secondly, enhance PPP in infrastructural development and thirdly, promote planned integrated development of towns and cities (GoIMUD, 2011b). The number of approved projects in this scheme was 787 by September 2011 at an aggregated value of USD 2.74 billion (UIDSSMT, 2011).

Investments in UIG and UIDSSMT for Urban Infrastructure
In total 1,323 projects have been approved by September 2011 with the aim to improve the urban infrastructure of numerous cities and towns across India. A comprehensive report undertaken by a High Powered Expert Committee (HPEC) under the leadership of the Ministry of Urban Development (GoIMUD, 2011c) calculated into which urban infrastructure sectors funding was released, according to Fig. 1 (as of 1 December 2010).

As Fig. 1 highlights, the largest shares of spending went to improving the water supply and road networks. This is little surprising as the GoIMUD (2011c) estimates that the investment over the two decades into India's urban infrastructure will be as high as INR 39.2 lakh crore (2009 prices) or USD 800 billion (4 November 2011 exchange rate). The following allocations of funds per sector would be as follows: USD 353 billion (INR 17.3 lakh crore) or 44 per cent on urban roads; improvements of water supply, sewerage, solid waste management and storm water drains would require USD 163 bullion (INR 8 lakh crore) or 20 per cent and the renewal of slums USD 81.5 billion (INR 4 lakh crore) or just 10 per cent of the total estimated budget.

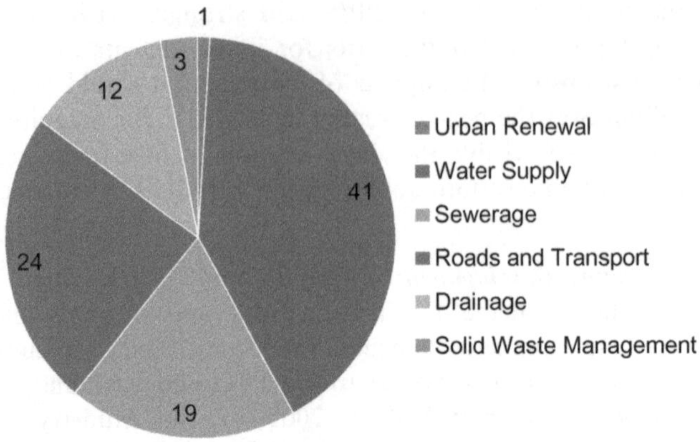

Fig. 1. Spending by Sector for UIG and UIDSSMT (in per cent). *Source*: GoIMUD (2011c).

Sub-Mission on Basic Services to the Urban Poor

Alongside the JNNURM and UIDSSMT, a third programme named Basic Services to the Urban Poor (BSUP) has been established as well in 2005 which includes the same cities as in the JNNURM (see Table 5), but is run by the Ministry of Housing and Urban Poverty Alleviation. Since the JNNURM and UIDSSMT programmes focus on issues related to urban infrastructure and local governance, the BSUP aims to improve the living conditions of the urban poor (Government of India Ministry of Housing and Urban Poverty Alleviation (GoIMHUPA), 2009). Accordingly, the key objectives of this sub-mission are as follows: firstly, provision of affordable housing, improved housing quality, water supply and sanitation services; secondly, provision of measures to secure linkages between asset creation and asset management to allow urban poor to become financially independent and thirdly, efforts should be increased to deliver civic amenities and utilities to improve the living conditions of urban poor (GoIMHUPA, 2009).

As of September 2011, 495 projects have been approved at an aggregated value of USD 5.84 billion (GoIMHUPA, 2011). However, the cost sharing in this scheme is done between three bodies, as shown in Table 6.

Therefore, the actual costs borne by the central government are only around half of the approved value. The remaining part is funded by the state and local governments. A key benefit of this funding scheme is that ULBs, local municipalities, gain ownership as they become part of the financing of projects.

Table 6. Indian Urban Sector Investment Requirement under BSUP Scheme.

Category	Grant Central Share	State/ULB/Parastatal Share, Including Beneficiary Contribution
Cities with over 4 million population	50%	50%
Cities with 1–4 million population	50%	50%
Cities/towns in North Eastern states and Jammu and Kashmir	90%	10%
Other cities	80%	20%

Source: GoIMHUPA (2009).

DISASTER RISK REDUCTION AT THE LOCAL LEVEL IN INDIA

The establishment of the JNNURM and the programmes running along with it are supportive elements to ensure that DRR takes place at the local level in urban areas. Furthermore, the 74th Constitutional Amendment (adopted in 1992) is the key legislative imperative for enhanced devolution of powers from the central government down to the local level. Hence, India's DMA has been established along this constitutional amendment (GoI, 2005).

As a result, the combination of key legal and financial requirements established over the past 6 years are promising steps towards inclusive DRR implementation in urban areas of India. However, the selective character of the UIG where the focus lies on only 63 identified cities bears some danger that other cities not part of the UIG become neglected. Thus, the UIDSSMT scheme may become a crucial financial instrument to allow those cities to get access to funding to improve their urban infrastructure.

As local level bodies, like cities and towns, have to follow the rigid hierarchical institutional system to get access to potential funding of projects this may in many cases be cumbersome and thus DRR as a new concept may not be prioritised to be integrated into development plans. This is a key problem of the new schemes/missions presented before; nonetheless, they go into the right direction to devolve more power to ULBs. Accordingly, the GoIMUD (2011c) calls for a New Improved JNNURM (NIJNNURM) which would eradicate the selective nature of the UIG and BSUP programmes under the JNNURM and replace the project-based character of the JNNURM by a more programme-based approach which would be open to any type of urban area. Furthermore, enhanced efforts are required to build and strengthen the capacities of the ULBs. Thus, government officers require more training and knowledge on how to run effectively local institutions. Finally, the devolution of power from central government to the state and local government under the 74th Constitutional Amendment provides ULBs with greater independence to govern with more autonomy. However, this devolution of more freedom and power to the local municipalities and towns is still lacking behind in progress due to a lack of reforms that would increase the financial resources and capacities of the ULBs (GoIMUD, 2011c). Thus, more money should be channelled to the ULBs from the central and state governments; additionally, tax reforms would alleviate the persistent financial problems of many cities and towns in India.

DISASTER RISK REDUCTION AND RESILIENCE IN URBAN AREAS OF INDIA

The JNNURM mission ought to be regarded as key step forward for inclusive DRR that has the potential to enhance the resilience of urban areas in India. As it not only addresses the many nuisances in India's cities related to their basic infrastructure but also pushes for the proper implementation of the 74th Constitutional Amendment that aims to improve the abilities and capacities of local bodies, like municipalities, to exert their daily duties (National Institute of Urban Affairs (NIUA), 2005). Although India's states are obliged to implement the 74th Constitutional Amendment, work from the NIUA (2005) emphasised that many states faced problems in dissolving executive powers to the municipalities; therefore, the JNNURM is welcomed to provide new incentives for states and municipalities to make use of their legal rights (JNNURM, 2011). The need for strengthened local institutions is very much in line with the international guidance on local level DRR provided by the KU (2010).

Despite the fact that the JNNURM does not explicitly declare its mission as one that is providing DRR in cities its objectives, however, are targeting the various climate disaster risks discussed in Chapter 2. From tackling urban poverty, improving the basic services, enhancing the institutional capacities, integrative (sustainable) planning and strengthening the autonomy of cities all aspects referred as stresses in Chapter 2 (Fig. 8) are to some extent targeted in the JNNURM. Therefore, this mission is providing DRR which essentially is making cities more resilient to all types of disasters.

SUMMARY

This chapter emphasised on the evolution of DRR from the international perspective and how it is addressed in the Indian context. Despite the inexistence of a clear policy on DRR, the GoI has laid out the legal and financial basis for inclusive DRR implementation for urban areas in India. The adoption of the DMA in 2005 and decisions of various funding schemes by the GoI are crucial to fulfil the expectations defined in the HFA. To conclude, in recent years, various efforts have been undertaken at the national level in India to provide legal and financial frameworks that have the potential to bring DRR in urban areas and thus make cities more resilient to disasters.

REFERENCES

Association for Southeast Asian Nations (ASEAN). (2009). *Roadmap for ASEAN community. ASEAN Secretariat, Jakarta.* Retrieved from http://www.aseansec.org/publications/RoadmapASEANCommunity.pdf. Accessed on November 20, 2011.

Brundtland, G. H. (1987). *Our common future. The World Commission on Environment and Development.* Oxford: Oxford University Press.

Government of Himachal Pradesh (GHP). (2011a). *Himachal Pradesh State Policy on Disaster Management. Department of Revenue, Shimla.* Retrieved from http://hpsdma.nic.in/NewsFiles/HPDMP2011.pdf. Accessed on October 19.

Government of Himachal Pradesh (GHP). (2011b). *State disaster management plan. GHP.* Retrieved from http://hpsdma.nic.in/DisasterManagement/HPSDMP.pdf. Accessed on October 19.

Government of India (GoI). (2005). *Disaster Management Act 2005. Ministry of Home Affairs.* New Delhi: Government of India.

Government of India (GoI). (2008). Environment and climate change. In *Government of India planning commission. Eleventh five year plan (2007–2012)* (Vol. 1, pp. 191–222). New Delhi: Government of India planning commission.

Government of India (GoI). (2011). *Draft – Faster, sustainable and more inclusive growth. An approach to the 12th five year plan.* New Delhi: Government of India Planning Commission.

Government of India Ministry of Housing and Urban Poverty Alleviation (GoIMHUPA). (2009). *Modified guidelines for submission on basic services to the urban poor (BSUP).*

GoIMHUPA. Retrieved from https://jnnurmmis.nic.in/jnnurm_hupa/jnnurm/BSUP%20revised%20guidelines%202009.pdf. Accessed on October 3, 2011.

Government of India Ministry of Housing and Urban Poverty Alleviation (GoIMHUPA). (2011). *Jawaharlal Nehru national urban renewal mission. Fund allocation and utilisation (BSUP/IHSDP)*. Retrieved from https://jnnurmmis.nic.in/jnnurm_hupa/jnnurm/DMU_REPORT_JNNURM.pdf. Accessed on October 31.

Government of India Ministry of Urban Development (GoIMUD). (2009). *Urban infrastructure development scheme for small & medium towns. GoIMUD*. Retrieved from http://urbanindia.nic.in/programme/ud/uidssmt_pdf/overview.pdf. Accessed on November 20, 2011.

Government of India Ministry of Urban Development (GoIMUD). (2011a). *Guidelines for Jawaharlal Nehru urban renewal mission*. New Delhi: GoIMUD.

Government of India Ministry of Urban Development (GoIMUD). (2011b). *Guidelines for urban infrastructure and development scheme for small & medium towns*. Retrieved from http://urbanindia.nic.in/programme/ud/uidssmt_guidelines.htm. Accessed on November 3.

Government of India Ministry of Urban Development (GoIMUD). (2011c). *Report on Indian urban infrastructure and services*. Retrieved from http://niua.org/projects/hpec/Final Report-hpec.pdf. Accessed on November 4.

Government of Tamil Nadu. (2008). *Government Order (G.O.) Ms. No. 689. Tamil Nadu Revenue Department, Chennai*. Retrieved from http://www.tn.gov.in/gorders/rev/rev_e_689_2008.pdf. Accessed on October 31, 2011.

Gujarat State Disaster Management Authority (GSDMA). (2003). *Gujarat State Disaster Management Act, GSDMA, Ahmedabad*. Retrieved from http://www.gsdma.org/dmact.pdf. Accessed on October 30, 2011.

International Decade for Natural Disaster Reduction (IDNDR). (1994). *Yokohama strategy and plan of action for a safer world. Guidelines for Natural Disaster Prevention, Preparedness and Mitigation*. Yokohama: IDNDR.

Jawaharlal Nehru National Urban Renewal Mission (JNNURM). (2009). *Overview. Government of India, Delhi*. Retrieved from http://jnnurm.nic.in/wp-content/uploads/2011/01/UIGOverview.pdf. Accessed on October 20, 2011.

Jawaharlal Nehru National Urban Renewal Mission (JNNURM). (2011). *Fund allocation and release (UIG)*. Retrieved from http://urbanindia.nic.in/DMU/JNNURM/DMU-JNNURM.pdf. Accessed on November 30.

Kyoto University (KU). (2010). *A guide for implementing the Hyogo Framework for Action by local stakeholders*. United Nations, Geneva. Consultation version.

National Disaster Management Authority (NDMA). (no dateA). *Annexure: List of states, districts and cities under drr programme*. NDMA, Delhi. Retrieved from http://ndma.gov.in/ndma/undpdrrpoject/listofstates.pdf. Accessed on September 30, 2011.

National Disaster Management Authority (NDMA). (no dateB). *Progress report. GoI-UNDP disaster risk reduction programme (2009–2011)*. NDMA, Delhi. Retrieved from http://ndma.gov.in/ndma/undpdrrpoject/progressreport.pdf. Accessed on September 30, 2011.

National Institute of Urban Affairs (NIUA). (2005). *Impact of the constitution (74th Amendment) act on the working of urban local bodies* (volume 1). Series 106. Retrieved from http://www.niua.org/research_studies_2006.asp. Accessed on October 19, 2011.

North Eastern Region Urban Development Programme (NERUDP). (no date). *Overview*. Retrieved from http://urbanindia.nic.in/programme/ud/nerudp_pdf/Overview.pdf. Accessed on November 21, 2011.

South Asian Association for Regional Cooperation (SAARC). (2007). *SAARC Development Goals (SDGs) (2007–2012), taking SDGs forward.* Retrieved from http://www.saarc-sec.org/uploads/publications/file/TAKING%20SDGs%20FORWARD%20(saarc-sec_20100616032736.pdf. Accessed on November 20, 2011.

Surjan, A. K. (2008). *Resilience to disaster & climate risk through community based environmental improvement in urban India.* Doctoral thesis, unpublished, Graduate School of Global Environmental Studies, Kyoto University, Kyoto.

United Nations (UN). (1992). *Report of the United Nations conference on environment and development.* A/CONF.151/26 (Vol. 1). Retrieved from http://www.un.org/documents/ga/conf151/aconf15126-1annex1.htm. Accessed on September 20, 2011.

United Nations. (1998). Kyoto protocol to the United Nations Framework Convention on Climate Change. UNFCCC. Retrieved from http://unfccc.int/resource/docs/convkp/kpeng.pdf. Accessed on November 30, 2011.

United Nations (UN). (2004). *Review of the Yokohama strategy and plan of action for a safer world.* A/CONF.206/L.1. Retrieved from http://www.unisdr.org/2005/wcdr/intergover/official-doc/L-docs/Yokohama-Strategy-English.pdf. Accessed on September 20, 2011.

United Nations Centre for Regional Development (UNCRD). (1989). *Challenges of the IDNDR.* UNCRD Meeting Report Series; no. 32. Retrieved from http://www.hyogo.uncrd.or.jp/publication/pdf/Proceedings/1989IntlSymposium.pdf. Accessed on November 21, 2011.

United Nations International Strategy for Disaster Reduction (UNISDR). (2007). *Words into action: A guide for implementing the Hyogo Framework.* Geneva: United Nations.

United Nations International Strategy for Disaster Reduction (UNISDR). (2009). *200 UNISDR terminology on disaster risk reduction.* Geneva: UNISDR.

United Nations International Strategy for Disaster Reduction (UNISDR). (2010). *A ten-point checklist for local governments – Ten essentials for making cities resilient.* Retrieved from http://www.unisdr.org/english/campaigns/campaign2010-2015/documents/230_tenpointchecklist.pdf. Accessed on September 20, 2011.

United Nations International Strategy for Disaster Reduction (UNISDR). (2011). *Global Assessment report on disaster risk reduction.* Geneva: United Nations.

United Nations International Strategy for Disaster Risk Reduction (UNISDR). (2005). *Hyogo Framework for 2005–2015: Building the resilience of nations and communities to disasters.* Retrieved from http://www.unisdr.org/wcdr/intergover/official-doc/L-docs/Hyogo-framework-for-action-english.pdf. Accessed on November 22, 2011.

Urban Infrastructure Development Scheme for Small & Medium Towns (UIDSSMT). (2011). *Fund allocation and utilization (UIDSSMT). A summary statement.* Retrieved from http://urbanindia.nic.in/DMU/UIDSSMT/DMU-UIDSSMT.pdf. Accessed on November 3, 2011.

Verstappen, H. (1993). The international decade for natural disaster reduction and the IGU flood hazard research programme. *Geojournal, 31*(4), 309–312.

CHAPTER 5

DEVELOPMENT AND APPLICATION OF A CLIMATE DISASTER RESILIENCE INDEX IN CHENNAI AND OTHER INDIAN CITIES

INTRODUCTION

In this and the following chapter, the concept of resilience, described in Chapter 3 and the need for DRR in Chapter 4, is applied to measure the climate-related disaster resilience of different types of urban areas, ranging from entire cities, zones to neighbourhoods. The key question to be answered in this chapter is: how to measure the climate-related disaster resilience of a city? In other words, how can different aspects of urban risks and impacts of climate change get linked to form a single tool capable of assessing the complexity of an urban area or city in relation resilience?

The structure of this chapter provides first an overview of the origin of the Climate Disaster Resilience Index (CDRI) methodology and its initial use. Secondly, the modified CDRI is described in detail. Following this methodological part, results from the CDRI applied among 13 Indian cities are presented, whereby a detailed study of Chennai emphasises on the application of the CDRI at the community level of a city.

Origin of the Climate Disaster Resilience Index

The origin of the CDRI goes back to 2007 when a team of various researchers in the laboratory of International Environment and Disaster

Management (IEDM) initiated the idea to quantitatively assess the resilience of cities to hydro-meteorological disasters, such as floods, storms, rainfall-induced landslides, etc. Thereby, the CDRI included five dimensions (physical, social, economic, institutional and natural). The aim was to measure the current capacity of different properties of an urban community, including the provision of basic urban services. However, the different parameters and variable comprising the five dimensions were rudimentary and unequal weight was given to each of the five dimensions. Meaning the physical dimension had stronger weight to measure aspects of basic urban services, described in Chapter 2.

Nonetheless, the initial methodology was applied in 15 cities in the Asian region through a partnership with CITYNET (regional network of local authorities for the management of human settlements in the Asia-Pacific region). Through this partnership local municipalities were requested to fill up a questionnaire about their city's condition. The results of the assessments in these 15 cities were published in 2009 in a publication named City Profile (Shaw, 2009). The participating cities in this initiative were: Banda Aceh, Bangkok, Colombo, Danang, Dhaka, Hanoi, Ho Chi Minh, Hue, Iloilo, Makati, Mumbai, San Fernando, Sukabumi, Suwon and Yokohoma.

CLIMATE DISASTER RESILIENCE INDEX/INITIATIVE – METHODOLOGY

Followed by this initial pilot study, extensive modifications of the CDRI were undertaken to improve its applicability to measure the climate-related disaster resilience of cities. In particular, elements from the HFA were integrated into the CDRI to reflect issues related to DRR.

Elements of the Climate Disaster Resilience Index

Based on explanations drawn in Chapter 2, to properly address issues in communities, different dimensions define whether a city can be considered as resilient. Thus, the physical, social, economic, institutional and natural dimensions aim to address to address the complexity of a city to function in case of a disaster.

Within these five dimensions, 25 parameters (five in each dimension), and 125 variables (five each parameter, 25 each dimension), cover key aspects of resilience to climate-related disasters, see Table 1. Based on an

Table 1. Dimensions, Parameters and Variables of CDRI.

Dimensions	Parameters and Variables
Physical	*Electricity* (access, availability, (interruption), supply capacity, alternative capacity) *Water* (access, availability, (interruption), supply capacity, alternative capacity) *Sanitation and solid waste disposal* (access to hygienic sanitation, collection of waste: treated, recycled) *Accessibility of roads* (transportation network, paved roads, accessibility during normal flooding, accessibility during catastrophic flooding, roadside covered drain) *Housing and land-use* (building code, buildings with non-permanent structure, buildings above water logging, ownership, population living in proximity to polluted industries)
Social	*Population* (population growth, population under 14, population informal settlers, (urban poor), population density at day and night) *Health* (population suffer from waterborne/vector-borne diseases, population suffer from waterborne diseases after a disaster, access to primary health facilities, capacity of health facilities during a disaster) *Education and awareness* (literacy rate, population's awareness about disasters, availability of public awareness programmes/disaster drills, access to internet, functionality of schools after disaster) *Social capital* (population participating in community activities/clubs, acceptance level of community leader (in ward), ability of communities to build consensus and to participate in city's decision-making process (level of democracy), level of ethnic segregation) *Community preparedness during a disaster* (preparedness (logistics, materials, and management), provision of shelter for affected people, support from NGOs/CBOs, population evacuating voluntarily, population participating in relief works)
Economic	*Income* (population below poverty line, number of income sources per household, income derived in informal sector, income disparity, per cent of households that have reduced income due to a disaster) *Employment* (formal sector: per cent of labour unemployed, per cent of youth unemployed, per cent of women employed, per cent of employees come from outside the city; per cent labour employed in informal sector) *Household assets* (households have: television, mobile phone, motorized vehicle, non-motorized vehicle, basic furniture) *Finance and savings* (availability of credit facility to prevent disaster, accessibility to credits, accessibility to credits for urban poor, saving practice of households, household's properties insured, catastrophic risk financing) *Budget and subsidy* (funding of DRM part of budget, budget for DRR sufficient, availability of subsidies/incentives for residents to rebuild houses, alternative livelihood, health care after a disaster)
Institutional	*Mainstreaming of DRR and CCA* (mainstreaming of CCA and DRR in city's land-use, housing, transport and environmental plans, mainstreaming of CCA and DRR in school education system)

Table 1. (*Continued*)

Dimensions	Parameters and Variables
	Effectiveness of city's crisis management framework (incorporation of disaster management plan including uncertainties of climate change, effectiveness of emergency team (leadership), efficiency of trained emergency workers during a disaster, existence of alternative decision-making personnel) *Effectiveness of city's institutions to respond to a disaster* (effectiveness of city's formal institutions during a disaster, effectiveness of city's informal organisations during a disaster, efficiency of trained emergency workers, frequency of training programmes for emergency workers, learning from disasters) *Institutional collaboration with other organisations and stakeholders* (city's dependency to external institutions/support, collaboration and interconnectedness with neighbouring cities, city's collaboration with national government, NGOs and private organisations during a disaster) *Good governance* (enforcement of disaster risk management, accountability of city government during a disaster, implementation of building codes, effectiveness of early-warning systems, existence of disaster drills)
Natural	*Intensity/severity of natural hazards* (floods, cyclones, rainfall-induced landslides, heat waves, droughts (water scarcity)) *Frequency of natural hazards* (floods, cyclones, rainfall-induced landslides, heat waves, droughts) *Ecosystem services* (quality of city's biodiversity, soils, air, water bodies, urban salinity) *Land-use in natural terms* (area vulnerable to climate-related hazards, urban morphology, settlements on hazardous ground, amount of Urban Green Space (UGS), loss of UGS) *Environmental policies* (existence and compliance rate to environmental policies, implementation of efficient waste management system (RRR), implementation of mitigation policies to reduce air pollution, food security after a disaster)

extensive literature review (Adger, 2000; Bruneau et al., 2003; Campanella, 2006; Cutter et al., 2008; Folke, 2006; Godschalk, 2003; Holling, 1973; Huq, Kovats, Reid, & Satterthwaite, 2007; Kadushin, 2004; McEntire, 2001; Rose, 2007; Rosenthal & Kouzmin, 1996; Satterthwaite, Huq, Pelling, Reid, & Romero-Lankao, 2007; Vale & Campanella, 2005; Wisner, Blaikie, Cannon, & Davis, 2004; World Bank, 2009) discussed in Chapters 3 and 4, different dimensions, parameters and variables were derived, as shown in Table 1, to define the resilience of communities in an urban system (city) to climate-related disasters. After the assessment of the available literature on resilience, communities in urban areas and DRR, the five dimensions were attributed with the same number of parameter and variables to comprehensively reflect aspects of an urban areas' resilience to climate-related disaster.

According to Chapter 2, various risk drivers, like aspects of urbanisation, urban poverty, declining quality and capacity of ecosystems, unplanned growth, etc., characterise many cities in developing countries. To alleviate these risk drivers, sustainable development is needed to connect different elements; however, in order to give a better overview about the relevance of each of the five dimensions based on the literature review, the below list summarises the characteristics of a resilient urban community:

- *Physical*: studies (Cannon, Twigg, & Rowell, 2003; Gaillard, Pangilinan, Cadag, & Le Masson, 2008; Twigg, 2007) on post-disaster livelihood assessments emphasise, for example, on the need for people to have secure electricity and water supply to recover quickly from a disaster. In other words, a solid physical infrastructure is crucial for urban areas to absorb a disaster and thus, apart from functioning urban services the built environment (e.g. houses) need to meet highest building and engineering standards.
- *Social*: various scholars (Cannon et al., 2003; Kadushin, 2004; Murphy, 2007; Paton, 2003) stress the beneficial support of strong social capital, social networks, disaster awareness among communities to not only withstand a disaster but also to better respond it. Furthermore, Tobin and Whiteford (2002) point out that intact and well-functioning health capacities (facilities, networks) during situations of disaster are imminent to reduce avoidable losses of human lives.
- *Economic*: Rose (2004, 2007) emphasises on the adequate allocation of financial resources and effective organisation of the economic sector to support and develop incentives to reduce losses from disasters. Available insurance schemes and financial systems would have the potential to provide pre- and after-disaster funding (public and private) which are beneficial to sustain a disaster from the economic perspective.
- *Institutional*: the mainstreaming of climate change adaptation (Trohanis, Shah, & Ranghieri, 2009) alongside effective emergency management (McEntire, 2001) are two aspects which require a strong institutional setup to ensure their implementation before, respectively, their functioning during a disaster. Furthermore, the ability of the institutions to cooperate with other stakeholders (communities, NGOs, private organisations, etc.) and to provide good governance is crucial for well-functioning authorities before, during and after a disaster.

- *Natural*: the protection of the natural environment (ecosystems, urban green space) is crucial to reduce the probability of disasters to occur and to uphold its coping capacity during times of disasters. Moreover, to ensure the safeguarding of the natural environment, the compliance to land-use and environmental policies forms part of the natural resilience of an urban area which is often challenged by intense use of all types of ecosystems (air, water, soil, etc.).

To conclude, the CDRI tries to disentangle specific aspects of risk drivers, like urban poverty, urbanisation, unplanned development or the quality of ecosystems, and measures them through different factors (parameters and variables) represented in various dimensions (see Table 1). Additionally, the focus is on the ability of different systems (dimensions) to alleviate the probability of shocks and to respond effectively if they occur (focus on social and institutional dimensions).

Calculating the CDRI

While the CDRI questionnaire consists of five dimensions (Table 1), it is further defined by another five parameters which are again represented by five variables measuring a parameter in more detail. As a result, 125 variables divided evenly into 25 parameters and five dimensions define the resilience of a particular urban system; whereby, each variable (x_1, x_2, …, x_5), allows five different choices between not available or poorly available (score 1) and best with a score of 5. In addition, a weighting scheme requires that variables within a parameter, consisting of five variables, have to be ranked (w_1, w_2, …, w_5) depending on their importance (low importance [1], high importance [5]) in shaping the final score of a particular parameter and resilience dimension. Because of this simple structured questionnaire with the uniform numbers for each parameter and variable ranging between one and five, it allows a transparent adoption of the formula named weighted mean (Fig. 1) to calculate the CDRI scores for each variable, parameter and dimension in a standardised and harmonised approach.

$$\frac{\sum_{i=1}^{n} w_i x_i}{\sum_{i=1}^{n} w_i} = \frac{w_1x_1 + w_2x_2 + w_3x_3 + w_4x_4 + w_5x_5}{w_1 + w_2 + w_3 + w_4 + w_5}$$

Fig. 1. Formula – Weighted Mean for Calculating a Score of a Parameter.

Data Collection Process and Limitations of the CDRI

The CDRI framework is developed to accommodate the context urban areas in India where little secondary data is neither available at the city level nor at lower administrative levels, such as zones or wards in the case of Chennai. Therefore, the questionnaire relies to some extent on experts from departments or local engineers of a city to provide responses to the CDRI questionnaire, according to the description above.

Thus there are some limitations of this method, for example, it is unlikely that the suggested parameters can cover the entire complexity of a community or urban system (King, 2001). Also, some parts of the CDRI questionnaire would provide better results, like in the social, physical and economic dimensions, if conducted at the household level rather than at the city or zone level in the case of Chennai (King & MacGregor, 2000), this aspect is further elaborated in Chapter 7. Furthermore, subjective responses from engineers in completing the questionnaires cannot be excluded, as not all variables are available in form of secondary data. Finally, the frequency and intensity of the occurring natural hazards, reflected in the first two parameters of the natural dimension, are measures of exposure (vulnerability) rather than resilience.

Nonetheless, the CDRI represents an approach to understand the functionality and condition of different sectors which shape the resilience of an urban system to climate-related disasters. Accordingly, it is an attempt to systematically define indicators which are expected to measure the resilience of an urban area. Rather than assessing the disaster resilience of urban areas in a qualitative approach, as undertaken by Tanner, Mitchell, Polack, and Guenther (2009) in Chennai and other Asian cities, the CDRI aims to map out the resilience based on parameters and variables in a quantitative way. As a result, the CDRI provides a rough picture (map) of the current condition of an urban area about its ability to respond to potential climate-related disasters.

CLIMATE DISASTER RESILIENCE OF INDIAN CITIES

As discussed in Chapter 2, urban areas of India are challenged by various risk drivers due to impacts of urbanisation and hydro-meteorological hazards. Thus, the CDRI assessment was applied to understand how a selected number of Indian cities are performing and whether they are capable to absorb climate-related disasters.

Scope of Study and Approach

In collaboration with the National Institute of Disaster Management (NIDM) of India, the Town and Country Planning Organisation and SEEDS India, a study was conducted between August and September 2009 to gain data from 12 selected cities (excluding Chennai). The CDRI data for Chennai was gained during a multi-city project (eight Asian cities) in 2010. Thereby, each municipality received a CDRI questionnaire, tailored to the Indian context, to fill-up after consulting different departments within their institutional body. Fig. 2 shows the geographical location of the cities.

Selected Cities and Data Collection

As Fig. 2 shows, the 13 selected cities have varying topographical and geographical characteristics ranging from river-side, coastal, mountainous, arid and mixed (combination of more than one geographical and topographical aspect, e.g. river-side and coastal). As mentioned before, the data from

Fig. 2. Location of Selected Cities for CDRI Study.

the 12 cities (excluding Chennai) was received through a questionnaire which was designed in partnership with the above organisations to fully reflect the Indian context. After designing and reviewing (among experts) the questionnaire was distributed through NIDM to an identified 72 cities based on size and aspect (see Fig. 2). The addressees of the questionnaire were Municipal Corporations who were requested to gather the data within their departments to provide responses to the CDRI questions. This data collection was undertaken during a two-month period which also included a workshop in Delhi, held in September 2009, to bring together representatives from the identified 72 cities.

Although the number of participants (21 cities) was higher than the final total sample size (12) from this project (Table 2), the workshop provided a platform for governmental officers to share their concerns and ideas on issues related to the climate-related disaster resilience of their cities (see Fig. 3). The information and insights gained from this event served to further understand the challenges that Indian cities are facing related to planning, climate disaster risk and climate change impacts.

In addition to the CDRI questionnaire and workshop, two field visits to Jaipur and Shimla (see Fig. 4) allowed getting a direct impression on the difficulties that the cities are faced due to climate disaster risk. In the case of Jaipur, as mentioned in Chapter 2, the shortage of water is often a problem if the monsoon is delayed in July. In contrast, Shimla is faced by problems related to rainfall-induced landslides due to its steep exposure which can be as high as 60 degrees. The picture on the bottom right in Fig. 4 shows a landslide that killed one person 1 year before the visit in 2009. This may not be very impressing; however, the central location of this landslide just below the main square of Shimla highlights the very high hazard risk pattern that exists across the city. Finally, both cities are facing unprecedented population growth rates beyond the Indian average in urban areas.

Table 2. Selected Cities for CDRI Study.

Size Aspect	River-Side	Coastal	Mountainous	Arid	Mixed
Small (up to 500,000)		Port Blair	Aizawl Shimla		
Medium (500,000 to 3 million)	Kanpur Varanasi			Amritsar Jaipur Nagpur	Bhubaneshwar Guwahati
Large (more than 3 million)	Delhi Kolkata	Chennai			

Fig. 3. Workshop at NIDM, Delhi, in September 2009.

Similar to Jaipur and Shimla, Delhi is faced by various climate disaster risk such as high population density particularly in the older part of the city, exhausted basic services (water, electricity and sewerage system) and high air pollution. Fig. 5 shows the old part of Delhi which has a potential density of up to 30,000 people per square kilometre – this is around five times as much compared to the average of Tokyo.

Results of CDRI Assessment

In the following sections, the results compare whether bigger cities have higher CDRI scores than smaller ones. Secondly, the aim is to understand whether the location of cities is having an impact on CDRI scores. Finally, the question is which dimensions, parameters and variables have lowest scores.

Results Based on Population Size of Cities

Table 3 shows the resilience scores of the 13 cities in relation to the population size, from the biggest to the smallest. The assumption that bigger cities

Fig. 4. Jaipur (Left) and Shimla (Right) in September 2009.

Fig. 5. Old Part of Delhi in September 2009.

Table 3. CDRI Scores (Overall and Dimension-Wise) Ranked Based on Cities' Population (2001).

City	Population in million	Physical	Social	Economic	Institutional	Natural	Overall
Delhi (R)	13.78	3.84	3.08	2.44	2.84	3.52	3.14
Kolkata (R)	4.57	4.16	3.68	2.42	3.48	3.40	3.43
Chennai (C)	4.34	2.92	4.08	3.06	3.56	2.83	3.29
Kanpur (R)	2.72	3.36	3.16	2.52	3.60	3.12	3.15
Jaipur (A)	2.32	4.04	3.32	2.44	2.76	3.08	3.13
Nagpur (A)	2.13	4.32	4.22	2.76	3.88	3.76	3.79
Varanasi (R)	1.20	2.99	3.14	2.52	2.58	3.08	2.86
Amritsar (A)	1.00	3.36	2.60	2.40	2.08	3.12	2.71
Guwahati (Mi)	0.82	3.68	3.52	2.44	3.04	4.07	3.35
Bhubaneshwar (Mi)	0.66	3.24	2.60	2.60	2.93	3.24	2.92
Aizawl (Mo)	0.23	3.16	4.24	2.24	2.36	2.56	2.91
Shimla (Mo)	0.14	3.44	3.44	2.52	2.20	2.19	2.76
Port Blair (C)	0.10	3.64	4.16	3.08	4.64	3.80	3.86

would have lower resilience scores due to higher demands on all the five dimensions cannot be confirmed, as there is no statistical significant correlation between the population size and dimension parameters.

Results Based on Location of Cities

As Fig. 6 shows, various cities are exposed by different types of hazard, for instance, cities located at the river-side of the Ganges, like Kanpur or Varanasi are susceptible to flooding events. Whereas mountainous cities are more prone to rainfall-induced landslides; accordingly, arid cities are likely affected by drought. Therefore, Table 4 puts the CDRI scores in relation to the location of the cities.

Although definite conclusions are difficult to draw due to the limited numbers of sample cities; however, the results show that the two mountainous cities, Aizawl and Shimla, have a particular low CDRI score for the natural dimension. Since both cities are located in hilly areas with steep slopes, they are particularly affected by regular yearly occurring rainfall-induced landslides which may explain why their natural CDRI score is relatively low compared to the other cities which are located in plain areas.

Results of Cities Mapped-Out

Figs. 6 and 7 show the results from the CDRI assessment mapped-out into spider diagrams. The detailed results for each dimension are as follows.

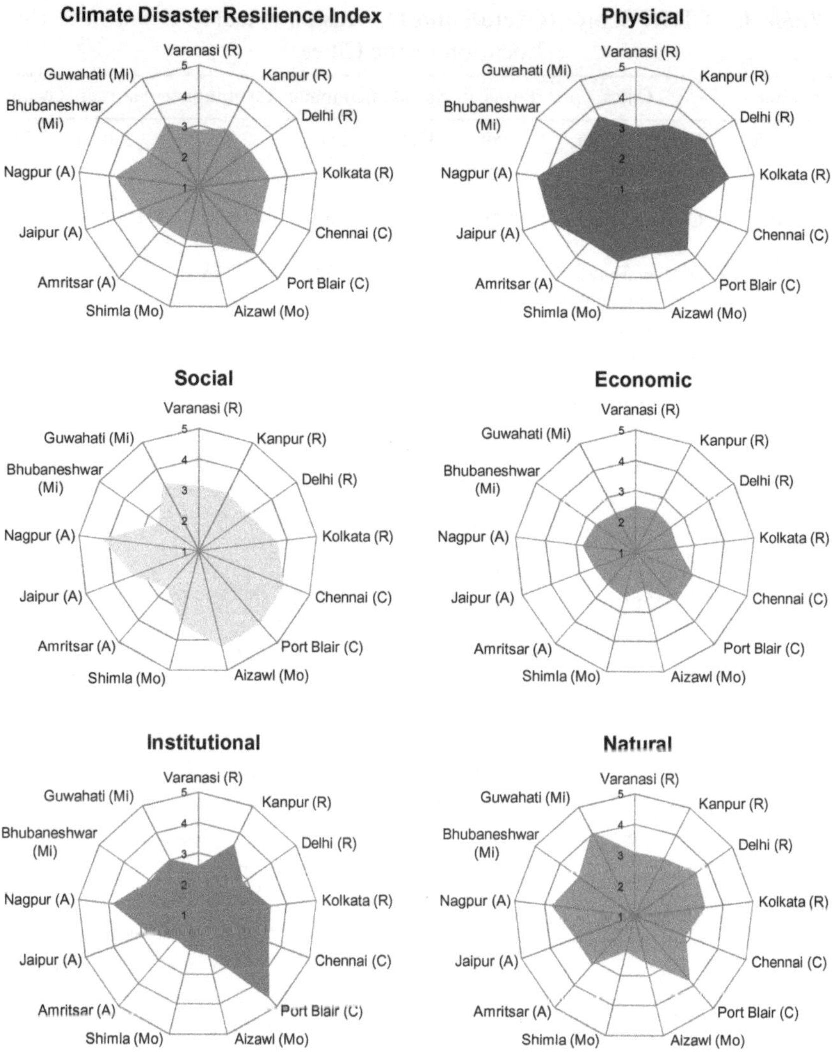

Fig. 6. CDRI Scores of All the Cities.

Physical. The average score among the five dimensions is highest (see Fig. 6) in the physical dimension compared to the other dimensions. The results also show little variations between the cities with rather equal average scores. However, scores between parameters of the physical dimension

Table 4. CDRI Scores (Overall and Dimension-Wise) in Relation to the Location of the Cities.

Location	Cities	Physical	Social	Economic	Institutional	Natural	Overall
River-side	Delhi, Kanpur, Kolkata, Varanasi	3.59	3.27	2.48	3.13	3.28	3.15
Coastal	Chennai, Port Blair	3.28	4.12	3.07	4.10	3.31	3.58
Mountainous	Aizawl, Shimla	3.30	3.84	2.38	2.28	2.37	2.83
Arid	Amritsar, Jaipur, Nagpur	3.91	3.38	2.53	2.91	3.32	3.21
Mixed	Bhubaneshwar, Guwahati	3.46	3.06	2.52	2.99	3.66	3.14

Fig. 7. CDRI Scores: Dimension- and Parameter-Wise.

show large variations; for example, the condition or accessibility of roads varies largely among the cities whereas issues related to housing and land-use show little variations. Another example, highlighting the need for detailed analysis, is Delhi which has very good accessibility of roads and highest provision of water, but below average sanitation and solid waste disposal.

The results from the physical dimension point out that parameter within the dimension do not correlate, for instance, higher scores in the parameter of sanitation and solid waste does not mean that the accessibility of roads is equally high. As a result, the CDRI scores of the 13 cities for the physical dimension have to be analysed independently between the cities, as the causes for the variations between the parameters is different based on the local context.

Social. Overall (Fig. 6) social resilience scores are varying between the cities. Furthermore, CDRI results from the social dimension also highlight large variations (Fig. 7) between parameters of health, education and awareness, social capital and community preparedness to disasters, but relatively little within the population parameter. The reason is likely that the impacts of urbanisation (population growth, urban poverty) are felt equally comparatively between the cities, leading to little variation. In contrast, the other four parameters show varying scores (Fig. 7) population parameter is represented by more quantitative data in contrast to social capital.

Economic. The scores of the economic dimension of the 12 cities are lowest among the five dimensions (Fig. 6). Looking at the shapes of the economic dimension (Fig. 7), there is little difference in the range of scores between the five parameters. Thus, a low economic score of a city is influenced by all five parameters. The reason for this generally low economic condition is due to limited availability for communities to generate income (low employment levels) which as a result, reduces their opportunities to accumulate wealth in form of household assets. Furthermore, as a consequence of this, they have reduced capacities to provide themselves and the government with money which could serve them before a disaster with financing protection measures and after the event for relief and rehabilitation activities.

Institutional. The CDRI scores in the institutional dimension show large variations between the 13 cities, ranging from very high to very low (Fig. 6). Good crisis management is followed by well-functioning institutional collaboration or mainstreaming of DRR and CCA into development plans

(Fig. 7). However, this large ambiguity between cities' institutional performances underlines that the CDRI approach is heavily challenged to understand institutional processes adequately. Thus, the results in the institutional dimension are difficult to interpret; however, the large variations between the cities are likely pointing out the different perceptions between the different local governments who were requested to undertake a self-evaluation of their governmental performance. Therefore, it must be concluded that results do not match the actual, for example, in the case of Port Blair the results show an almost perfectly functioning authority.

Natural. The natural dimension shows mixed results between the 12 cities. However, cities in the mountainous areas, such as Shimla and Aizawl, are having lower natural resilience compared to the cities located in less hilly areas (Fig. 6). The CDRI does not confirm the assumption that intense land-use leads to a lower quality of a city's ecosystem, it rather shows the opposite. Furthermore, the large variations between the parameters, that exemplifies the different natural conditions between the selected 12 cities.

Results Based on Characteristics of Dimensions and Parameters

In a statistical attempt, the results from the 13 cities' five dimension are correlated against each other. Accordingly, Table 5 shows that only few dimensions correlate significantly; however, correlation coefficient between the economic and institutional dimension demonstrate a statistically high score of $r = 0.79$. This means that the higher the economic score of a city the higher the institutional value. Hence, higher availability of economic resources for communities and the municipality may support the functioning of the institutional governance before and during a potential disaster. This example points out a certain dominance of economic-related aspects. Results, shown in the results part, indicate similar trends that the CDRI method is to some extent influenced by aspects of livelihood of communities. In other words, more wealth or financial capacities is beneficial to increase the value of the other dimensions.

Results Based on the 25 Parameters

The ranking between the average scores of the 25 parameters (Table 6) shows that economic parameters are among those with lowest scores. Surprisingly, aspects of urban infrastructure, such as accessibility of roads, water, electricity or sanitation and solid waste, are measured as comparatively high resilient.

Table 5. Correlation Coefficients between the Five Dimensions.

Physical and Social	Physical and Economic	Physical and Institutional	Physical and Natural	Social and Economic	Social and Institutional	Social and Natural	Economic and Institutional	Economic and Natural	Institutional and Natural
0.38	0.20	0.46	0.49	0.31	0.50	0.14	**0.79**	0.45	0.66

Table 6. Average CDRI Scores (Parameter-Wise) for All the Cities.

Parameter	Highest to Lowest
Housing and land-use	3.8
Health	3.73
Accessibility of roads	3.7
Water	3.63
Intensity of natural hazards	3.59
Education and awareness	3.55
Community preparedness	3.47
Electricity	3.45
Sanitation and solid waste	3.43
Frequency of natural hazards	3.41
Ecosystem services	3.37
Environmental policies	3.29
Social capital	3.28
Good governance	3.24
Knowledge dissemination	3.2
Institutional collaborations	3.2
Population	3.15
Crisis management	2.9
Income	2.75
Land-use	2.7
Household assets	2.68
Mainstreaming of DRR and CCA	2.63
Budget and subsidy	2.58
Finance and savings	2.38
Employment	2.28

Implications for Indian Cities

The CDRI assessment of the 13 Indian cities revealed several findings, for example, the ongoing urbanisation trend in many Indian cities is reflected in the social dimension where the population parameter tends to be lower than other parameters within this dimension (see Fig. 6). Also, the impacts of urbanisation are seen in the relatively low score of the land-use parameter (see Fig. 7). As this parameter reflects the loss of urban green space, and intensity of land-use, it confirms that urbanisation is actually taking place; however, to assume that the ecosystem would be harmed due to this phenomenon cannot be confirmed.

Regarding the economic dimension (Fig. 6), the CDRI results highlight that this dimension has lowest CDRI scores within the five dimensions.

The reason for this lies in the relatively low average per capita income of India compared to other countries in the Asian region – the CDRI for India was designed based on available literature which assumed higher overall economic conditions for Indian cities. Although, urban areas in India have higher average incomes than rural areas, these incomes are still lower than in other countries, like the Philippines or Thailand. Therefore, the CDRI score for the economic dimension does not mean that no economic growth in the cities is occurring, but rather exemplifies that the level of economic wealth is still relatively low in India. This example highlights further that the CDRI assessment is favourably applied among cities with similar cultural and legal conditions. Thus, results require to be interpreted based on the local context in qualitative approach.

Looking at the natural dimension, cities in mountainous areas (see Fig. 6) are particularly affected by frequently occurring rainfall-induced landslides, as shown by Aizawl and Shimla; thus, the frequently occurring natural hazards are challenging the resilience levels of the cities; however, looking at the social dimension, the results are above average indicating that citizens are prepared to face disasters. This underpins that high vulnerability (physical, natural, economic) and occurrence of climate-related disasters are not associated with social aspects of a city.

Summary

In this primary assessment about the resilience of 13 Indian cities, the concept of resilience applied a set of 125 variables, equally divided into five parameters, to measure the ability of cities to manage climate-related disasters. Despite the challenges raised on the Indian urban infrastructure in Chapter 2, the CDRI assessment does not fully confirm this notion. However, it cannot be ruled-out that the CDRI questionnaire at this stage (in 2009) was not yet fully calibrated to disclose the actual conditions of each dimension properly. Thus, the modified CDRI assessment at the zone level in Chennai (see next section) is a new evolved version taking into account lessons learned from this initial assessment.

Nevertheless, the CDRI applied on the 13 Indian cities provided an attempt to better understand how different aspects of a city influence the resilience to climate-related disasters.

CLIMATE DISASTER RESILIENCE OF CHENNAI

Introduction

In this part, the CDRI, described above, is modified to be applicable at the zone level of Chennai. The selection of Chennai as a study location is ideal to investigate aspects of urbanisation in combination with projected climate change. The following paragraphs emphasise on the reason for selecting Chennai as the primary study area for this doctoral research.

Firstly, Chennai has grown rapidly starting after the independence of India from the British Empire in 1947 and although population growth rates have stabilised in the past decade problems of urban poverty, exhausted urban services and unequal social characteristics are just a few of the earlier mentioned risks that are challenging the functioning of the city due to impacts of urbanisation.

Secondly, based on projections from Aggarwal and Lal (2001) the sea level is likely to rise by up to 49 cm in the 21st century which means increased potential of storm surges due to the relatively low altitude with just a few metres above sea level and coastal exposition of the city. Furthermore, potential changes in the occurrence, length (time period) and intensity (cyclones) of the post-monsoon period (October–December) may cause climate-related risk.

Thirdly, Chennai is experiencing unprecedented growth of its urban area (settlement) and its economy which has led to an expansion of the city area beyond the current size of 176 km^2 in October 2011. Although the research undertaken in these studies focuses on the 'old' layout of the city (zone and ward distribution), the potential impact of this urban expansion is expected to require sound planning to which the presented research methods and applications may provide contribution. Further details of this urban expansion are discussed in Chapter 8.

Fourth, the availability of a pro-active city government which provided logistical support, interest (launched a Safer Chennai Campaign in 2010, further discussed in Chapter 8), information (primary and secondary data) and direct access to the different study sites proved to be essential for conducting this type of research.

Fifth, the established relationship (MoU) between the University of Madras and Kyoto University provided local academic support which was crucial in supporting all parts of this research from a scientific and logistical point of view.

As a result, Chennai provides a perfect setting to analyse the climate-related disaster resilience of different neighbourhoods (zones) within the city. Knowing about variations in resilience levels has the potential to provide baseline information supporting the process of taking action at the planning level (discussed in Chapter 8).

Characteristics of Chennai

Chennai is located at the coast of the Bay of Bengal. Its current area size is 176 km^2 with a population density of around 26,597 people per square kilometre. The population size as of 2011 was 4.68 million (not including the metropolitan area).

The geographic location is at 13°5′2″N 80°16′12″E in the south-east of India (see Fig. 8). The city's climate is considered as tropical with dry (March–May) and wet (October–December) periods. The city is further characterised by two rivers which lie in a near 90 degree angle towards the sea. The northern river is named Cooum and the southern one is the Adyar River. Chennai as well as its outskirts lie in a former lagoon area which has low altitude (few metres above sea level) and also large parts of this area

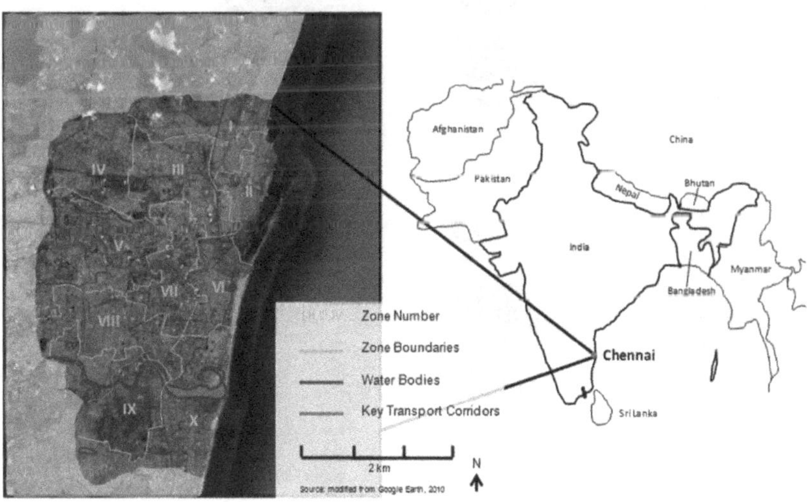

Fig. 8. Location of Chennai, India. *Source*: Map on the left is modified from Google Earth.

are marsh land with naturally high ground water levels. Thus, many parts of the city especially along canals are flood-prone areas.

History of Chennai's Development

Chennai or Madras was founded in 1639 by British businessmen belonging to the East India Company and with the establishment of the Fort St. George in the same year it became a seat of power at the Coromandel Coast (Muthiah, 2008). Following the foundation of the former Madras, today referred as Chennai, various villages scattered a few kilometres around Fort St. George began to grow steadily (Fig. 9) before the city amalgamated to a single unit in the middle of the 20th century (Muthiah, 2008).

The city's coastal location always favoured the development of sea trade-related industries and activities. Thus, the first piers at the location of today's port were established in 1861 allowing ships to harbour and stimulate sea trade (Muthiah, 2008). The port was established and expanded in the succeeding years and contributed to Madras' growth in both terms,

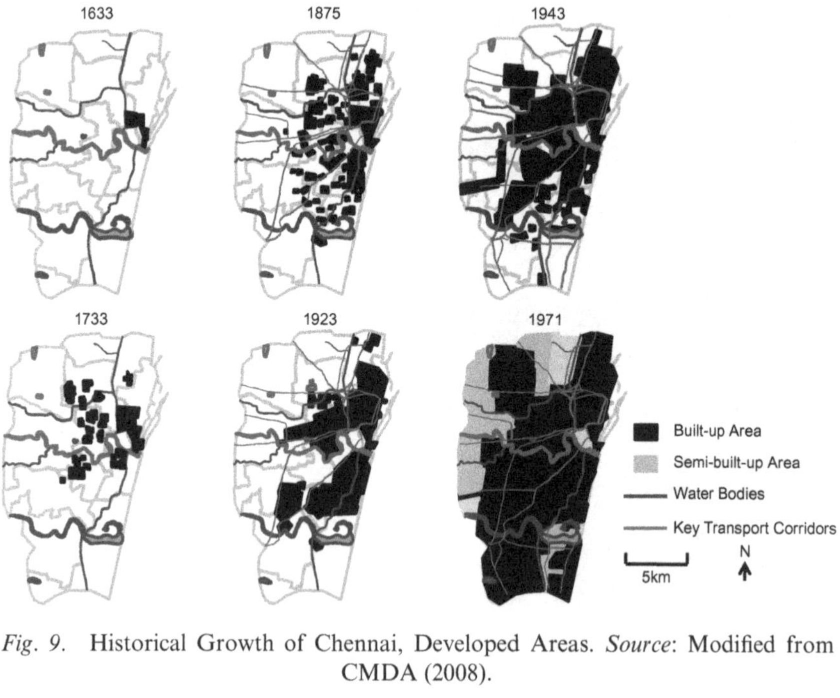

Fig. 9. Historical Growth of Chennai, Developed Areas. *Source*: Modified from CMDA (2008).

population-wise and also in economic terms (Muthiah, 2008). In 1901, the city area stretched over an area of 70 km^2 and had around 540,000 inhabitants. Because of high population growth rates of annually 5–6 per cent in the following decades, the city became in 1941 a provincial metropolis and an administrative and commercial centre (Chennai Metropolitan Development Authority (CMDA), 2008).

The population growth and city expansion continued in the subsequent years passing the million population mark in 1943. This population growth kept going after India gained independence in 1947 and became a Republic in 1950. Although economic growth increased sharply during this period until 1971, the downsides of this growth became visible as slums began to mushroom along the canals and water supply and drainage systems deteriorated in their quality.

Since 1971, the city's size of 176 km^2 includes 10 administrative zones (grey boundaries in Fig. 9) which carry out the civic needs of the residents. The municipality is referred as the Corporation of Chennai which manages the institutional functioning of the city. It is the oldest municipality in India formed in 1688 and its headquarters are the Ripon Buildings, established in 1913 in Victorian style.

Although the city's annual growth rate was around 1.72 per cent (see Table 7) during 1971 to 2001, this growth has slowed down in the past decade to 0.75 per cent per year (Census of India, 2011). The past growth of 1.72 per cent is close to the current world annual urban population growth rate of 1.76 per cent (United Nations Population Division (UNPD), 2010). The original location of the city's foundation, Fort St. George, is located in Zone II (Table 7), the old parts further include Zones III, VI and VII which are bordering (see Fig. 10 for overview of zones). The old parts were largely populated already before the middle of the 20th century. Therefore, not surprisingly, they had considerable lower population growth rates during 1971 to 2001.

In average, the population growth in the old parts (four zones) was 0.9 per cent per year compared to 2.54 per cent in the other six zones. As a result, it is expected that the population growth which took place in the past decade largely occurred in the outer areas of the city.

The population growth in absolute numbers was 337,442 from 2001 to 2011 or an increase of only 7.75 per cent. This number is low compared to the population growth which is recorded outside the city's boundary. For example, the Kancheepuram District bordering in the south and south-west of Chennai grew by 38.69 per cent during the same period of time. Thus, Chennai's population is stabilising as new areas outside the city are

Table 7. Demographic and Land-Use Condition in the 10 Zones of Chennai.

Area	Size in km^2	% Population Growth p.a., 1971–2001	Population Density in 2001	Key Characteristics
Zone I	17.3	2.40	23,699	Urban fringe: residential and industrial area
Zone II	11.52	0.15	32,638	Old part: industrial (port), commercial, institutional, and residential area
Zone III	13.51	1.07	34,048	Old part: large urban poor areas, residential and commercial area
Zone IV	19.76	2.72	25,151	Urban fringe: residential area
Zone V	26.38	3.08	20,545	Urban fringe: fast growing residential and commercial developments
Zone VI	10.15	0.15	33,694	Old part: institutional area, beach
Zone VII	12.9	0.60	26,976	Old part: commercial area
Zone VIII	13	2.12	35,846	Urban fringe: residential area
Zone IX	23.56	2.66	17,614	Urban fringe: residential, industrial, and institutional area, large park (greenspace)
Zone X	27.92	2.25	17,478	Urban fringe: fast developing area (commercial and residential)
City	(Total) **176**	(AVG.) **1.72203**	(AVG.) **26,768.9**	

Source: Modified from CMDA (2008).

growing instead. Although the population growth was only 0.75 per cent per year in Chennai, in areas located at the border of Chennai, annual population growth rates go, for example, as high as 16 per cent in case of Virugambakkam.

Looking at projections for the wider metropolitan area, Chennai is expected to become nearly a megacity by 2025 with a projected population of 9.9 million (UNPD, 2010). The metropolitan region includes three districts, Chennai, Thiruvallur and Kancheepuram, which together have a total area size of 1,189 km^2. The size of the urbanised population was around 7.34 million. As not only the Kancheepuram is growing rapidly but also Thiruvallur District grew by 35.25 per cent during 2001–2011

Fig. 10. Overview of Selected Zones in Chennai.

(Census of India, 2011), it is only question of time until the Chennai Metropolitan Area (CMA) trespasses the 10 million mark. The urban expansion of Chennai and issues related to the CMA are further discussed in Chapter 8.

To summarise this part briefly, Chennai City has grown rapidly over the past decades, but has now a stabilising population. It is considered that rapid urbanisation to some extent triggered the problems and challenges that Chennai faces today, which are discussed in the next section.

Urban Risks in Chennai

As mentioned before, during the 1950s until today the population and economic growth of the city rose steadily, but was not well planned or well designed. As a result, unplanned growth took place due to a lack of regulations (CMDA, 2008). This phenomenon persists until today; particularly, in areas along the outer parts of the city which were only developed recently, meaning over the past four to five decades. This unplanned development is further accelerated by poor urban governance (lack of enforcement of laws; e.g. building code) and widely existing corruption.

Another prevalent issue in Chennai is its large share of population who live in slums (see examples of slums in Figs. 11 and 12). According to official sources, in 2001, 820,000 people or 18.9 per cent of the total population were living in slums (CMDA, 2008). However, as of today, based on new counts in 2011, the city has around 1,674,018 of people living in a total of 799 slums, according to the Corporation of Chennai. This means the

Fig. 11. Slum along Adyar River.

Fig. 12. Urban Risk Areas in Chennai.

share of urban poor population augmented to 35.75 per cent of the total population or in other words, the slum population almost doubled!

Although this increase is sharp, a programme named Basic Services for Urban Poor (BSUP) is now in place to provide basic services for at least 422 identified slums. This scheme has provided social housing, electricity and water supply to urban poor people and has considerably improved the situation in many slums.

Despite the high rate of urban poor, there is a considerable number of growing middle class which is attracted to move to Chennai due to its flourishing IT and car manufacturing sectors. In particular, the southern and western parts just outside of the Chennai city area are witnessing the development of new headquarters of companies and business parks (CMDA, 2008). Therefore, the measured per capita income in 2001 of INR 21,738 or USD 440 at today's currency rate, which was two times higher than the Indian average at that time, is likely to have increased. According to a newspaper report referring to estimations by economists, Chennai's average per capita income is expected to rise up to USD 1,149 by 2015 (The Hindu, 2007). Projections from McKinsey (2010) for the year 2030 expect the yearly per capita income to be around USD 6,600 which would be roughly 15 times more compared to 2001. The development of various economies, like IT, car companies or other business parks, is expected to further grow along the urban fringe and thus, continue accelerating the economic growth of Chennai and its metropolitan area over the next decades (CMDA, 2008). To summarise briefly, the economic growth of Chennai is expected to contribute positively to the livelihood of the people in Chennai and therefore reducing prevalent urban risks. Although a large share of the entire population may not benefit from this growth equally, as the large number of urban poor indicate.

The downsides of the urbanisation are also visible in deteriorating ecosystems. As Chennai used to be once a popular destination for tourists to enjoy the lagoon type of environment during the 19th until early 20th century (Muthiah, 2008), little is left of this likable atmosphere. Coelho and Raman (2010) emphasise on the poor water quality of Chennai's waterways. As the city is characterised by a large number of canals and rivers, they are widely used as drainage systems to take up untreated household waste. Until recently even parts of industrial waste were routed directly into these waterbodies.

While not only the sewerage is reaching the waterbodies the city is challenged by a serious waste collection problem. Fig. 13 shows the sheer amount of waste dumped into the waterways. The impacts of this waste problem in Chennai is described in detail by Dahiya (2003) who blames

Fig. 13. Adyar River (Left) and Canal (Right) in Ward 131 as Waste Deposits.

the local government for inaction and little sense to treat this issue as serious. The unavoidable consequences of this waste problem lead to groundwater being polluted. Hence, the same water that is later being used for drinking purposes.

The high metal concentration of many aquifers in and around Chennai have been observed since many years and are of particular concern due to the fact that many dwellers residing near the waterbodies extract their drinking water from wells (hand pump) located in (see Fig. 14) or near their actual home (Somasundaram, Ravindran, & Tellam, 1993).

While the pollution of the groundwater is one issue, the high concentration of solid waste in certain parts of rivers and canals increase the risk for flooding as the water cannot drain as it gets blocked. Although the Corporation of Chennai has a solid waste management system in place, not in all areas the household waste is collected, particularly in areas where less fortunate people, as the results show from a case study, described in Chapter 7.

Even though the groundwater table has again risen in recent years, after too much water was extracted (CMDA, 2008), thanks to better water management and new sources from lakes outside of the city, the provision is still restricted to only 3–5 hours per day. As the entire water management is handled by the Chennai Metropolitan Water Supply and Sewerage Board (CMWSSB), it controls the supply of water to be discharged only during the night usually starting from midnight up to 5 am where water is pumped and routed into the city's pipeline. As a result, most households have rooftop tanks installed to store the amount of water needed per day. Again, obviously, richer people may have likely have better availability of such rooftop tanks compared to poorer people. This issue is again discussed in the case study presented in Chapter 7.

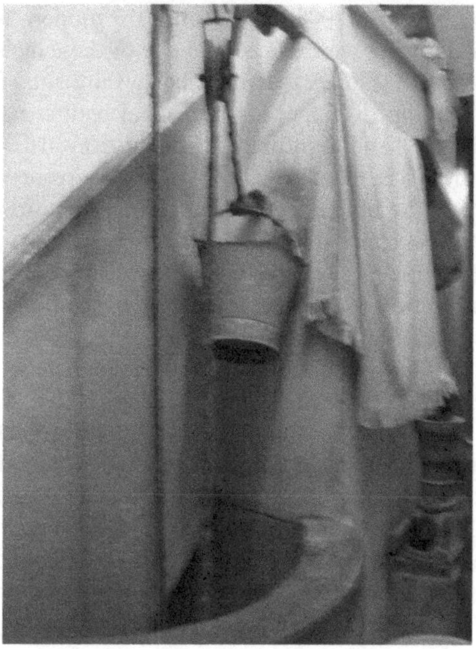

Fig. 14. In-House Water Well in Jafferkhanpet, Chennai.

Another identified urban risk is the provision of electric supply which is often interrupted especially during the hot and dry period of the year during the months from March to June. Thus, supply is always cut between 3 and 4 am and during the day blackouts are recorded frequently in all parts of the city. The city's electric supply is derived from two big coal-fired thermal power plants out of which one is located in Zone II.

To summarise shortly, Chennai is challenged by various urban risks that can trigger or aggravate disaster situations. Before turning the eye to the CDRI zone assessment, the potential threats of climate change in Chennai is elaborated.

Impacts of Climate Change-Related Disasters

The primary climatic threat to Chennai is the occasional occurrence of cyclones which can hit the city mostly during the post-monsoon period between October and December (Drescher et al., 2007; Revi, 2008). Followed by such cyclones, or sometimes before the advent of them, are heavy rainfalls leading to flooding. For example, in the years of 2005 before cyclone Fanoos and after Cyclone Nisha in 2008 and Jal in 2010 numerous casualties were counted and large damages were left behind due to these incidents. In Fig. 15, the trajectories of past cyclones that hit Chennai are shown. Usually, these cyclones start to develop over the Bay of Bengal before making land-fall at the coast of Tamil Nadu and eventually drifting northwards.

According to the IMD E-Atlas, 28 cyclones were recorded between 1959 and 2008. Although the effects of climate change, like more frequent and more severe natural hazards (IPPC, 2007) are not yet significant in this region, an expected increase is likely to have dramatic consequences on Chennai; especially, because of its low-lying exposition with just a few metres above sea level (Revi, 2008).

As mentioned in Chapter 2, sea-level rise is predicted to increase by at least 0.49 m in the 21st century (Aggarwal & Lal, 2001). As the occurrence of heavy rainfall events mostly due to cyclones and projected sea-level rise are the main natural hazards in Chennai related to climate change, other types such as heat waves are another risk factor that is rising in frequency (see Fig. 16). Although there is not a high statistically significant increase in the number of hot days ($T > 40$ degree Celsius), extended periods of hot temperatures near 40 degree Celsius are being suffered more often especially from April to June (dry and hot period), according to perceptions from local people.

The occurrence of droughts is seldom in Chennai; however, water shortages during the dry period occurred frequently in the past and until

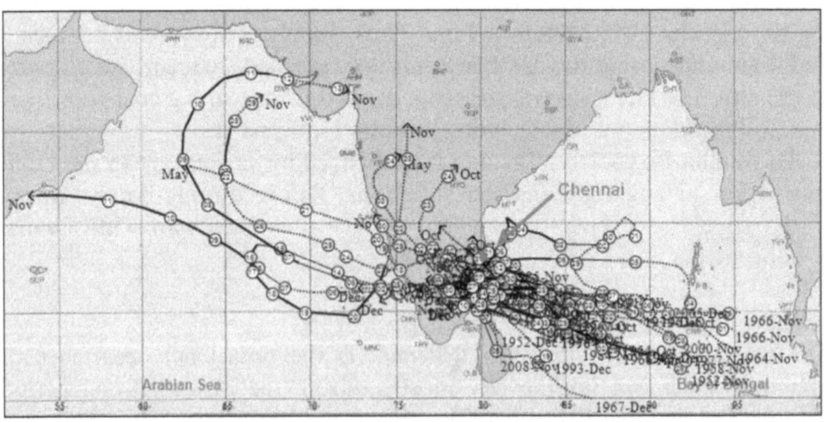

Fig. 15. Trajectories of Past Cyclones from 1959 to 2008 in North Part of Tamil Nadu State. *Source*: Adapted from IMD.

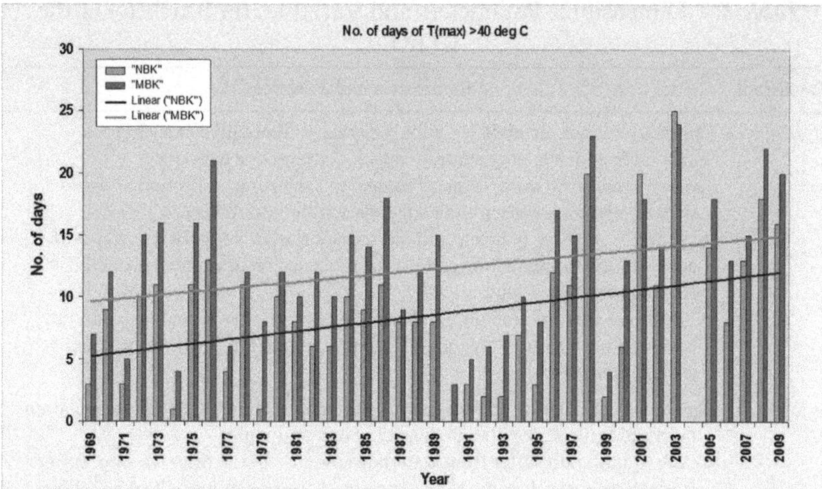

Fig. 16. Recorded Heat Waves in Chennai at Two Locations. *Source*: IMD.

today challenge the sufficient supply of water for Chennai's residents, as mentioned in the section 'Urban risks in Chennai'.

To conclude shortly, Chennai's history suggests that the combined power of stresses (risk drivers) and shocks (climate-related hazards) are likely to challenge the city's performance more severely if hit by a disaster; therefore, the CDRI shall reveal in what condition exactly the different zones of the city of Chennai are at the moment.

CLIMATE DISASTER RESILIENCE AT THE ZONE LEVEL OF CHENNAI

Methodology and Scope of Study

In order to understand to what extent Chennai is capable to absorb, manage and recover from climate-related disasters, a modified version of the CDRI undertaken for the Indian cities, shown above, is developed to address the context of the 10 zones in Chennai. Accordingly, the role or purpose of the zones as administrative units is reflected, particularly in variables related to the institutional dimension (see Table 8). As the zones are sub-units from the central administration, they deal, for example, with

Table 8. Dimensions, Parameters and Variables (in Brackets) of the CDRI.

Dimensions	Parameters and Variables
Physical	*Electricity* (access, availability, supply capacity, alternative capacity) *Water* (access, availability, supply capacity, alternative capacity) *Sanitation and solid waste disposal* (access to sanitation, collection of waste: treated, recycled, collection of solid waste after a disaster) *Accessibility of roads* (per cent of land transportation network, paved roads, accessibility during flooding, status of interruption after intense rainfall, roadside covered drain) *Housing and land-use* (building code, buildings with non-permanent structure, buildings above water logging, ownership, population living in proximity to polluted industries)
Social	*Population* (population growth, population under 14 and above 64, population informal settlers, population density at day and night) *Health* (population suffer from waterborne/vector-borne diseases, pop. suffer from waterborne diseases after a disaster, access to primary health facilities, capacity of health facilities during a disaster) *Education and awareness* (literacy rate, population's awareness about disasters, availability of public awareness programmes/disaster drills, access to internet, functionality of schools after disaster) *Social capital* (population participating in community activities/clubs, acceptance level of community leader (in ward), ability of communities to build consensus and to participate in city's decision-making process (level of democracy), level of ethnic segregation) *Community preparedness during a disaster* (preparedness (logistics, materials, and management), provision of shelter for affected people, support from NGOs/CBOs, population evacuating voluntarily, population participating in relief works)
Economic	*Income* (population below poverty line, number of income sources per household, income derived in informal sector, per cent of households have reduced income due to a disaster) *Employment* (formal sector: per cent of labour unemployed, per cent of youth unemployed, per cent of women employed, per cent of employees come from outside the city; per cent of child labour in zone) *Household assets* (households have: television, mobile phone, motorized vehicle, non-motorized vehicle, basic furniture) *Finance and savings* (availability of credit facility to prevent disaster, accessibility to credits, accessibility to credits for urban poor, saving practice of households, household's properties insured) *Budget and subsidy* (funding of DRM, budget for DRR sufficient, availability of subsidies/incentives for residents to rebuild houses, alternative livelihood, health care after a disaster)
Institutional	*Mainstreaming of DRR and CCA* (mainstreaming of CCA and DRR in: zone's development plans, ability (manpower) and capacity (technical) to produce development plans, extent of community participation in development plan preparation process, implementation of disaster management plan)

Table 8. *(Continued)*

Dimensions	Parameters and Variables
	Effectiveness of zone's crisis management framework (existence and effectiveness of an emergency team during a disaster: leadership, availability of evacuation centres, efficiency of trained emergency workers during a disaster, existence of alternative decision-making personnel) *Knowledge dissemination and management* (effectiveness to learn from previous disasters, availability of disaster training programmes for emergency workers, existence of disaster awareness programmes for communities, capacity (books, leaflets, etc.) to disseminate disaster awareness programmes (disaster education), extent of community satisfaction from disaster awareness programmes) *Institutional collaboration with other organisations and stakeholders, during a disaster* (zone's dependency to external institutions/support, collaboration and interconnectedness with neighbouring zones, zone's cooperation (support) with central corporation department for emergency management, cooperation zone's ward officials for emergency management, zone's institutional collaboration with NGOs and private organisations) *Good governance* (effectiveness of early warning systems, existence of disaster drills, promptness of zone body to disseminate emergency information during a disaster to communities and transparency of zone body to disseminate accurate emergency, capability of zone body to lead recovery process)
Natural	*Intensity/severity of natural hazards* (floods, cyclones, heat waves, droughts (water scarcity), tornados) *Frequency of natural hazards* (floods, cyclones, heat waves, droughts (water scarcity), tornados) *Ecosystem services* (quality of city's: biodiversity, soils, air, water bodies, urban salinity) *Land-use in natural terms* (area vulnerable to climate-related hazards, urban morphology, settlements on hazardous ground, amount of Urban Green Space (UGS), loss of UGS) *Environmental policies* (use of zone level hazard maps in development activities, extent of environmental conservation regulations reflected in development plans, extent of implementation of environmental conservation policies, implementation of efficient waste management system (RRR), implementation of mitigation policies to reduce air pollution)

issues related to collecting taxes, planning approvals, certificates (birth) and many others related to administrative support to their residents. Therefore, the selection of these zones was seen as ideal to get a more detailed insight into the local context through institutional collaboration.

The CDRI assessment was undertaken between January and February 2010 in collaboration with the Public Works Department of the Corporation of Chennai. Through local governmental support Assistant Executive Engineers (AEE) working in each of the 10 zones were approached to provide information and responses to the CDRI.

The role of these AEEs is to carry out civic work (see Fig. 17) apart from their responsibility to lead their zones during disaster situations. As their daily works requires them to move around in their zones to check the provision of basic services and also to collect grievances from local residents, they are close to the pulse of their zones' population. Therefore, they have considerable knowledge about the local situation and problems.

Following the data collection, a workshop (see Fig. 18) was held from 18 to 19 February 2010 where all the zone AEEs were invited alongside with other officers from governmental organisations, like CMWSSB or Tamil Nadu Electricity Board to deepen get further insights on the challenges the city faces related to the functioning of basic services as well as the situation within the different neighbourhoods. Furthermore, primary activities were undertaken to develop a Climate Action Plan (CAP) based on the perceptions from governmental officers (see further details about CAP in Chapter 8).

Results from the Climate Disaster Resilience Assessment of the 10 Zones

Overall Resilience

The overall resilience of Chennai is varying between the 10 zones (see Fig. 19) and points out that areas with higher economic development,

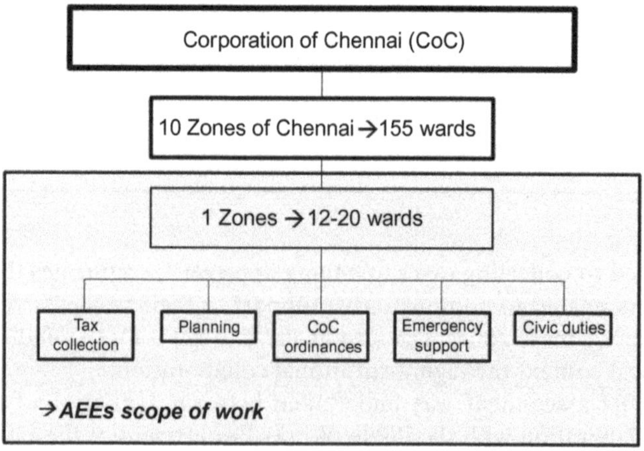

Fig. 17. AEEs Scope of Work.

Fig. 18. Local Climate and Disaster Resilience in Chennai Workshop (Top Left), Mayor of Corporation of Chennai (Second from Right), Workshop Activities with AEEs (Bottom).

lower population density and better environmental condition have higher resilience levels. Particularly, the northern areas are weaker compared to the growing southern and western parts of the city. The key findings from the overall resilience of the 10 zones give an early indication that resilience is likely linked to issues related to livelihood conditions. However, before going into depth about how the results from the CDRI assessment ought to be understood, each dimensions' resilience is summarised.

Physical Resilience

The physical resilience in Chennai is lower in the inner parts (Zones II, III, VI and VII) of the city. This area also forms the historical part (see Fig. 7) of Madras and has high population densities (>30,000 people per square kilometre) (see Table 7) demanding high provisions of basic services, such as the provision of water or electricity, and also the collection of solid waste and maintenance of roads. In contrast, the newly developed areas

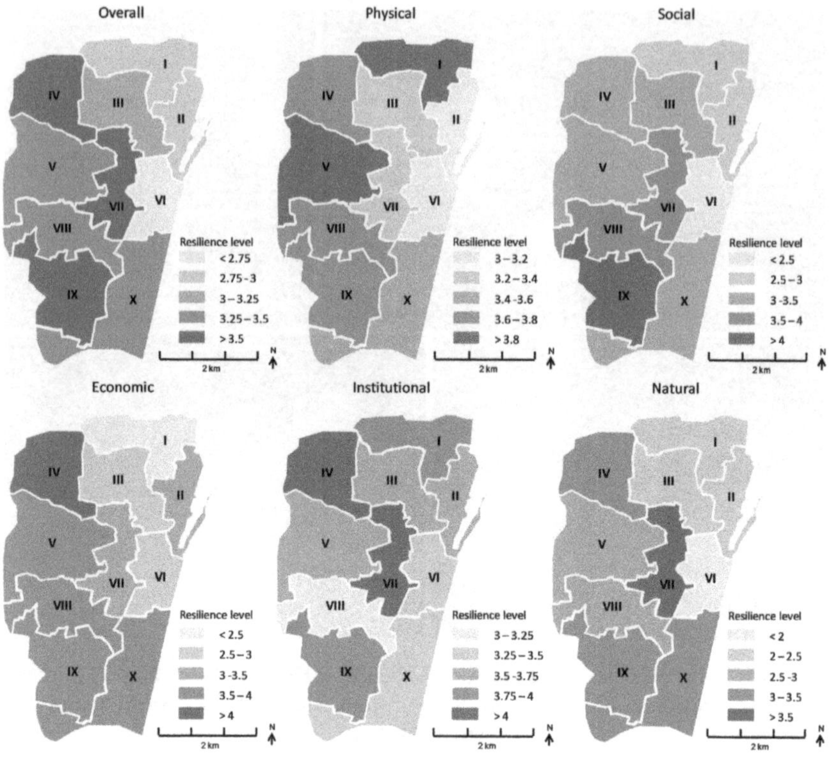

Fig. 19. Results from CDRI Assessment (Overall and All the Five Dimensions).

along the outer parts of the city have better infrastructure which is reflected in higher physical CDRI scores.

Areas with high physical resilience are those which were only recently established, meaning in the past three to four decades. In contrast, the physical infrastructures in the inner and older parts seem to be neglected and not properly maintained. This situation is not surprising since most of the attention from public or semi-private organisations, providing basic services, is devoted to increasing capacities in areas outside of the core areas to connect new people. Interestingly, this focus on developing new capacities and neglecting already existing installations is not only the case for the above mentioned organisations, as private housing developers often follow the same path to fully re-invest gained money from a sold development complex (e.g. house and office building) into new projects instead of conserving the existing housing infrastructure. As a result, maintenance is neglected causing overall the physical infrastructure of the city to slowly deteriorate, particularly in the older parts of the city.

The above interpretations of lower physical resilience in the older parts of the city are supported by statistical analysis where the average annual population growth is correlated to the results from the CDRI assessment of the physical dimensions (see Fig. 20). The result shows, based on Pearson correlation technique, that these two variables have a correlation coefficient of $r = 0.96$!

Although this high correlation result supports the earlier conclusions gained through the CDRI assessment it, however, denies assumptions raised by various scholars (Pelling, 2003; Wisner et al., 2004) that areas growing rapidly have lower quality of physical infrastructures. Based on the correlation between population growth and the CDRI scores of the physical resilience in Chennai these assumptions cannot be supported. In contrary, higher development activity ultimately leads to state-of-the art or modern constructions.

A possible reason for abandoning developed areas may lie in the structure of funding for urban development projects in India. As described in

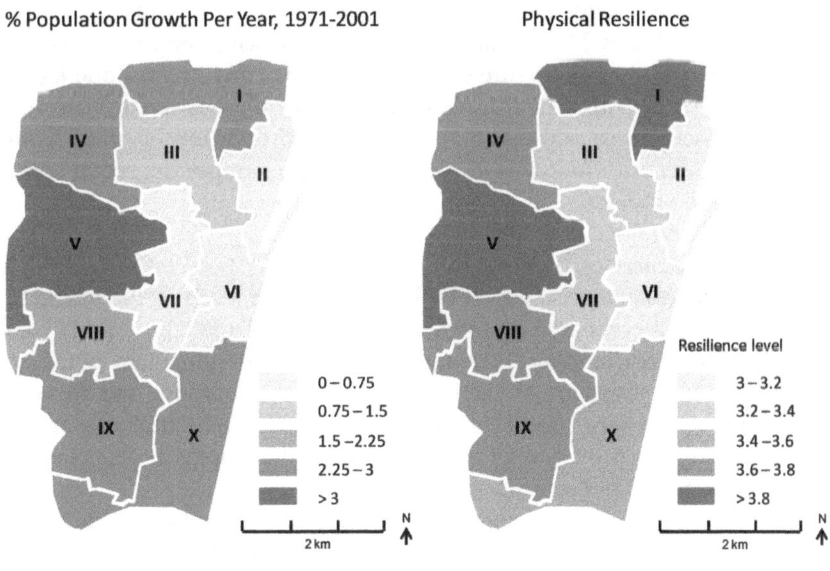

Fig. 20. Per cent of Population Growth per Year (1971–2001) versus Physical Resilience.

Chapter 4, major parts of funding if not all, for example, from the JNNURM scheme go into new development projects rather than conserving the existing infrastructures. Furthermore, the current Indian administrative system where in general local municipalities have limited financial resources compared to the state or the central government make it difficult to fund urban renewal projects which are in their nature not very attractive for politicians to add to their credentials. These issues are discussed further in Chapter 8.

Social Resilience

In contrary to the overall and physical results, the social resilience does not show a pattern related to trends of urbanisation. However, Zones II and VI and areas in the north of Chennai have lower resilience, due to larger proportions of urban poor living in these areas which also contribute to tensions among communities. The high proportion of urban poor in these areas lower the scores for most of the social parameters defined in the CDRI. Therefore, health, education and awareness to climate-related disasters are lower, as well as their social capital which is based on issues related to trust and civil engagement. Harriss (2005) and Coelho and Venkat (2009) discuss the lack of strong civil societies in Chennai. Although, not highly significant, trends show from case study results in Chapter 7 that communities in Ward 79 in Zone VI had lower participation in local social activities compared to comparable communities in Ward 131 in Zone IX.

The reasons for this situation cannot be uniformly answered, however, based on observations and discussions, the high proportion of urban poor and related insecurity in low resilient communities hamper the social cohesion among the residents.

Economic Resilience

The economic resilience is shaped by lower resilience levels in the northern parts of Chennai compared to the south and west of the city. Apart from the flourishing port in Zone II, comparatively little development is taking place in the northern part of the city; unlike in the southern and western parts of the city, where large IT centres and car companies are being established. The higher provision of employment in these areas is reflected by higher economic CDRI scores.

Although the economic resilience map resembles the overall and physical CDRI results, the commercial centre in the old part of the city in Zone VII denies a strong correlation. However, the low economic resilience of Chennai's northern areas points out that urbanisation is, in general, favourable in economic terms as new industries establish and jobs are created. Accordingly, wealth is generated leading to higher accession of household assets and saving opportunities of residents. These are all parameters reflecting the economic CDRI.

To find proper responses to this unequal development in Chennai is difficult; however, the fact that the international airport of Chennai is located in the south-west has attracted large investments for development projects in recent years. The significance of the airport may not be underestimated as a key economic driver. Since IT and international companies may chose headquarters that are in close proximity to a gateway leading them to other parts of India and international destinations. Thus, the absence of key facilities that would attract investments in the north of Chennai compounds this trend of unequal economic prosperity.

Institutional Resilience

The institutional resilience between the 10 zones in Chennai does not vary much. This is largely due to the administrative purpose of the zones to carry out work for the public which is directed in a top–down manner by the headquarters of the Corporation of Chennai. Because of this administrative system, the range of initiating or deviating from governmental orders is limited for the zones. In other words, zones have little autonomy to come up with new ideas that would allow them to initiate activities that would make them institutionally more resilient. For example, the decision to develop a local disaster management plan is taken either by the Councillors, representing wards in Chennai, or even comes from higher state-level authorities.

As a consequence, it is little surprising that institutional resilience scores deviate from each other. In contrary, it shows that governmental orders are equally carried throughout the city, which must be regarded as a positive trend. Thus, areas with a less fortunate economic situation are not less equipped or supported to carry out the civic services at the local level.

Natural Resilience

The northern areas have lower natural resilience than areas in the centre or south of the city. This is largely because of heavy polluting industries being located in Zones I, II and III, such as the port (Zone II), a large waste collection site (Zone I) and a big coal-fired power plant (between Zones I, II and III). As all these three zones including Zone VI are characterised by various canals and river systems, they directly suffer from the fact that industries and household discharge waste into these waterways. As a result, ecosystems (water, biodiversity, etc.) are in poor conditions.

On the other hand, the central part of Chennai (Zone VII) where little industrial activity is taking place and which is rather shaped by residential and commercial activities is having the highest natural resilience among all the zones; although, one of the key transport corridors of the city is lining through this area. Accordingly, this highest score needs to be put into perspective that the variables reflecting the parameters of land-use in natural terms or environmental policies are less relevant in the central part of the city. For example, the loss of urban greenspace took already place before the time horizon of 50 years as it was already developed long time ago. Moreover, the fact that little industries are located in this area and it is largely resided by middle or upper middle class residents and also many commercial (offices) companies are located there has comparatively positive impacts on the natural resilience of this area.

Other Findings

The comparative study among the 10 zones of Chennai allows correlating different dimensions and parameters together. As shown earlier, correlations were found between population growth and the physical resilience. In this section, all the values from the 25 parameters have been correlated against each other; however, only correlation coefficients which scored higher than $r = 0.8$ are listed. These results support to some extent aspects of livelihood (income-related aspects), for example, the linkage between higher income leading to higher accession of household or enhanced demand to land (Income and land-use in natural terms). As a result, the CDRI reveals the inclusion of aspects (livelihood) that have direct impacts on the ability of individuals (people) to build capacities that would prevent them from potential disasters. The income-related results further suggests that continuing economic growth is likely to benefit a large majority (middle and higher-income groups) of Chennai's population.

Another crucial result is the linkage between social capital and environmental policies and also the relationship between aspects of education and community preparedness. This has particular potential in the process of enhancing the implementation of environmental policies in the city. Additionally, the focus ought to be put on so-called soft-related issues which would make use of the capacities and collective power of communities, in particular the middle-class, in enhancing Chennai's resilience to climate-related disasters. This finding is also one of the reason why the CAP presented in Chapter 8 will focus more on action measures related to soft adaptation measures.

The results from Table 9 also highlight the need to build stronger capacities of local governmental staff. The relationship, for example, between a functioning crisis management and good governance supports the objectives raised by the JNNURM-related programmes that enhanced emphasis is required on educating and providing knowledge to governmental officer. This would strengthen their ability to carry out work that is going beyond providing the minimal or essential governmental services to Chennai's residents.

Results from the CDRI assessment allow the correlation not only between parameters but also between dimensions. Hence, the natural and social dimensions show the highest correlation significant correlation with a coefficient of $r = 0.84$ (see Fig. 21) among the different combinations correlating the dimensions.

This result emphasises that higher awareness about disasters and the environment lead to better protected (environmentally) areas. Thus, areas where the education level is low and communities have little cohesion are more likely to pay less attention to the needs to conserve the natural resources.

As the CDRI assessment required each parameter to be weighed based on the local context regarding its importance to influence the resilience level, Table 10 lists the parameters from high to low. Accordingly, sanitation and solid waste and community preparedness are weighed as those

Table 9. Correlations between Different Parameters of the CDRI, City-Wide (Chennai).

Interpretation	Above 0.80	Correlation Coefficient
○	Income and land-use in natural terms	0.90
⊕	Social capital and environmental policies	0.88
○	Income and household assets	0.86
○	Income and ecosystem services	0.83
○	Household assets and ecosystem services	0.83
⊕	Education and awareness and community preparedness	0.81
△	Mainstreaming and knowledge dissemination	0.81
△	Crisis Management and good governance	0.81
○	Income and finance and savings	0.80

○ mainly income-related; ⊕ socio-ecological-related; △ governance-related.

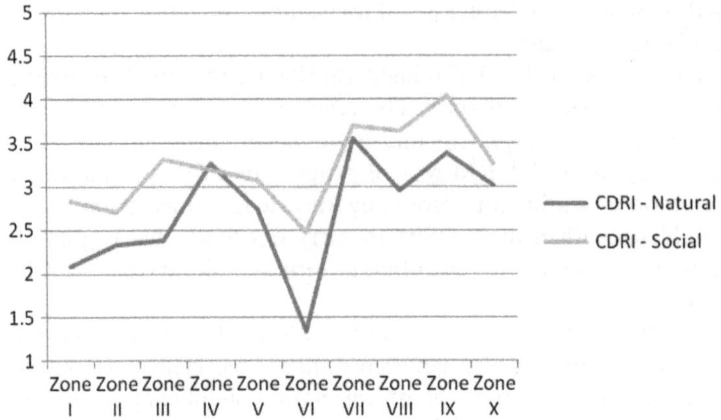

Fig. 21. Correlation between Natural and Social CDRI Dimensions.

Table 10. Weighting of Most Important Parameters to Influence Resilience Levels.

Parameters: Average Weighting of Parameters, Highest to Lowest	Weight
Sanitation and solid waste	4.2
Community preparedness	4.2
Water	3.8
Environmental policies	3.6
Mainstreaming of DRR and CCA	3.5
Crisis management	3.5
Budget and subsidy	3.4
Income	3.3
Education and awareness	3.2
Land-use	3.2
Finance and savings	3.1
Frequency of natural hazards	3.1
Employment	2.9
Good governance	2.8
Health	2.7
Electricity	2.6
Knowledge dissemination	2.6
Institutional collaborations	2.6
Ecosystem services	2.6
Social capital	2.5
Intensity/severity of natural hazards	2.5
Population	2.4
Accessibility of roads	2.3
Household assets	2.3
Housing and land-use	2.2

parameters that are most relevant to shape the ability of the communities to respond to climate-related disasters. However, this list also highlights the sectors where local government officials perceive the highest problems of their area. Therefore, it is not surprising that water or the implementation of environmental policies, DRR and CCA are regarded as top priorities since these sectors are currently challenged (water, quality of environment) or have not yet been incorporated into planning schemes (DRR and CCA).

Limitations of Results

Although absolute values are calculated for measuring the different dimensions of the 10 zones, the scope of interpretation has to be limited to comparing the 10 zones. Thus, a high score in one dimension in a particular zone does not necessarily imply that it is exempted from any further actions improving their resilience. Instead this local CDRI assessment focuses on the quantitative comparison between the zones with a similar cultural and political context.

Summary

This chapter focused on the application of the CDRI at the zone level of Chennai and thereby, revealed different resilience levels between different parts of the city. The CDRI results emphasised various connection between dimensions and parameters which suggest orientating the focus on enhancing Chennai's resilience to climate-related disasters in particular through soft adaptation-related actions. These findings are crucial to support the formulation of a CAP for Chennai which is further discussed in Chapter 8.

REFERENCES

Adger, W. N. (2000). Social and ecological resilience: Are they related? *Progress in Human Geography*, *24*(3), 347–364.

Aggarwal, D., & Lal, M. (2001). *Vulnerability of Indian coastline to sea-level rise*. New Delhi: Indian Institute of Technology. Centre for atmospheric sciences.

Bruneau, M., Chang, S. E., Eguchi, R. T., Lee, G. C., O'Rourke, T. D., Reinhorn, A. M., … Von Winterfeldt, D. (2003). A framework to quantitaively and enhance the seismic resilience of communities. *Earthquake Spectra*, *19*(4), 733–752.

Campanella, T. J. (2006). Urban resilience and the recovery of New Orleans. *Journal of the American Planning Association*, *72*(2), 141–146.

Cannon, T., Twigg, J., & Rowell, J. (2003). *Social vulnerability, sustainable livelihoods and disasters. conflict and humanitarian assistance department and sustainable livelihoods support office*. London: Department for International Development.

Census of India. (2011). *Provisional population totals paper 1: Tamil Nadu*. Government of India. Retrieved from http://censusindia.gov.in/2011-prov-results/data_files/tamilnadu/3.Tamil%20Nadu_PPT_2011-BOOK%20FINAL.pdf. Accessed on October 30.

Chennai Metropolitan Development Authority (CMDA). (2008). *Second master plan for Chennai metropolitan area, 2026*. Chennai: CMDA.

Coelho, K., & Raman, N. V. (2010). Salvaging and scapegoating: Slum evictions on Chennai's waterways. *Economic & Political Weekly*, *155*(21), 19–23.

Coelho, K., & Venkat, T. (2009). The politics of civil society: Neighbourhood associationism in Chennai. *Economic & Political Weekly*, *64*(26–27), 358–367.

Cutter, S. L., Barnes, L., Berry, M., Burton, C., Evans, E., Tate, E., & Webb, J. (2008). A place based model for understanding community resilience to natural disasters. *Global Environmental Change*, *18*, 598–606.

Dahiya, B. (2003). Peri-urban environments and community driven development. *Chennai, India. Cities*, *20*(5), 341–352.

Drescher, A., Glaser, R., Pfeiffer, C., Vencatesan, J., Schliermann-Kraus, E., Glaser, S., … Dostal, P. (2007). *Risk assessment of extreme precipitation in the coastal areas of Chennai as an element of catastrophe prevention. 8*. Forum DKKV/CEDIM: Disaster Reduction in Climate Change, October 15–16, Karlsruhe.

Folke, C. (2006). Resilience: The emergence of a perspective for social-ecological system analyses. *Global Environmental Change*, *16*, 253–267.

Gaillard, J., Pangilinan, M. R. M., Cadag, J. R., & Le Masson, V. (2008). Living with increasing floods: Insights from a rural Philippine community. *Disaster Prevention and Management*, *17*(3), 383–395.

Godschalk, D. R. (2003). Urban hazard mitigation: Creating resilient cities. *Natural Hazards Review*, *4*(3), 136–143.

Harriss, J. (2005). *Middle class activism and poor people's politics: An exploration of civil society in Chennai*. Working Paper Series No. 05-72. Development Studies Institute, London School of Economics and Political Science, London.

Holling, C. S. (1973). Resilience and stability of ecological systems. *Annual Review of Ecological Systems*, *4*, 1–23.

Huq, S., Kovats, S., Reid, H., & Satterthwaite, D. (2007). Editorial: Reducing risks to cities from disasters and climate change. *Environment and Urbanization*, *19*(1), 3–15.

Intergovernmental Panel on Climate Change (IPCC). (2007). Summary for policymakers. In S. Solomon, D. Qin, M. Manning, Z. Chen, M. Marquis, K. B. Averyt, … H. L. Miller (Eds.), *Climate change 2007: The physical science basis. Contribution of Working Group I to the Fourth Assessment Report of the Intergovernmental Panel on Climate Change*. Cambridge: Cambridge University Press.

Kadushin, C. (2004). Too much investment in social capital? *Social Networks*, *26*, 75–90.

King, D. (2001). Uses and limitations of socioeconomic indicators of community vulnerability to natural hazards: Data and disasters in northern Australia. *Natural Hazards*, *24*, 147–156.

King, D., & MacGregor, C. (2000). Using social indicators to measure community vulnerability to natural hazards. *Australian Journal of Emergency Management*, *15*(3), 52–57.

McEntire, D. A. (2001). Triggering agents, vulnerabilities and disaster reduction: Towards a holistic paradigm. *Disaster Prevention and Management*, *10*(3), 189–196.

McKinsey. (2010). India's urban awakening: Building inclusive cities, sustaining economic growth. McKinsey Global Institute, McKinsey & Company.

Murphy, B. L. (2007). Locating social capital in resilient community-level emergency management. *Natural Hazards*, *41*, 297–315.

Muthiah, S. (2008). *Madras rediscovered*. Chennai: Westland Limited.

Paton, D. (2003). Disaster preparedness: Social-cognitive perspective. *Disaster Prevention and Management*, *12*(3), 210–216.

Pelling, M. (2003). *The vulnerability of cities: Natural disasters and social resilience*. London: Earthscan.

Revi, A. (2008). Climate change risk: An adaptation and mitigation agenda for Indian cities. *Environment and Urbanization*, *20*(1), 207–229.

Rose, A. (2004). Defining and measuring economic resilience to disasters. *Disaster Prevention and Management*, *13*(4), 307–314.

Rose, A. (2007). Economic resilience to natural and man-made disasters: Multidisciplinary origins and contextual dimensions. *Environmental Hazards*, *7*, 383–398.

Rosenthal, U., & Kouzmin, A. (1996). Crisis management and institutional resilience. *Journal of Contingencies and Crisis Management*, *4*(3), 119–124.

Satterthwaite, D., Huq, S., Pelling, M., Reid, H., & Romero-Lankao, P. (2007). *Adapting to climate change in urban areas: The possibilities and constraints in low- and middle-income countries*. Human Settlements Climate Change and Cities Discussion Series 1. London: IIED.

Shaw, R. (2009). *City profile*. Graduate School of Global Environmental Studies, Kyoto University. Retrieved from http://www.preventionweb.net/english/professional/publications/v.php?id = 8168. Accessed on September 20, 2010.

Somasundaram, M. V., Ravindran, G., & Tellam, J. H. (1993). Ground-water pollution of the Madras urban aquifer, India. *Ground Water*, *31*(1), 4–11.

Tanner, T., Mitchell, T., Polack, E., & Guenther, B. (2009). *Urban governance for adaptation: Assessing climate change resilience in ten Asian cities*. Brighton: Institute of Development Studies.

The Hindu. (2007, January 5). Economist predicts reduction in poverty ratio by 2012. *The Hindu*. Retrieved from http://www.hindu.com/2007/01/05/stories/2007010518901000.htm. Accessed on November 30, 2011.

Tobin, G. A., & Whiteford, L. A. (2002). Community resilience and volcano hazard: The eruption of Tungurahua and evacuation of the faldas in Ecuador. *Disasters*, *26*(1), 28–48.

Trohanis, Z., Shah, F., & Ranghieri, F. (2009). *Building climate and disaster resilience into city planning and management processes*. Washington, DC: The World Bank. Sustainable Development Department East Asia and the Pacific Region.

Twigg, J. (2007). *Characteristics of a disaster-resilient community: A guidance note*. Disaster Risk Reduction Interagency Coordination Group, DFID: Benfield.

United Nations Population Division (UNPD). (2010). *World urbanization prospects: The 2009 Revision*. New York, NY: United Nations. United Nations Population Division, Department of Economic and Social Affairs.

Vale, L., & Campanella, T. (2005). *The resilient city: How modern cities recover from disaster*. New York, NY: Oxford University Press.

Wisner, B., Blaikie, P., Cannon, T., & Davis, I. (2004). *At risk: Natural hazards. People's vulnerability and disasters*. Cambridge, MA: MIT Press.

World Bank. (2009). *Climate resilient cities: A primer on reducing vulnerabilities to disasters*. Washington, DC: The World Bank.

CHAPTER 6

PERCEPTIONS OF COMMUNITY LEADERS ABOUT ENHANCING THE CLIMATE DISASTER RESILIENCE OF COMMUNITIES IN CHENNAI, INDIA

INTRODUCTION

Following the CDRI assessment the key question is how to address the different gaps and problems related to Chennai's resilience to climate-related disasters and also how to increase the city's resilience. For this, an Action-oriented Resilience Assessment (AoRA) has been developed as a subsequent tool, following the CDRI, to examine the responsibilities of different identified stakeholders (local government, communities, academia, private organisations and NGOs) in the process of implementing defined actions aimed at enhancing the resilience of Chennai to disasters.

ACTION-ORIENTED RESILIENCE ASSESSMENT

AoRA Methodology

The need for sound planning of communities is particularly important in an urban area like Chennai which is characterised by pressures due to urbanisation. As the past decades have shown in Chennai (see Chapter 5), the city experienced a dramatic increase of people which caused problems, such as exhausted urban infrastructure, urban poverty and deteriorating ecosystems which require decisive actions for improvement. However, since spatial planning in India and in general in developing countries is conducted usually with little involvement of the public in a top-down attitude (King, 2008), the views, knowledge and ideas of communities may not receive sufficient attention in the formulation of plans. Moreover, the potential of communities to exert collective power is often neglected. Similarly, the management of disasters and emergencies often relies on decisions taken by the local authorities with little consultation among communities in developing countries (King, 2008) which is disputed since local people are the first respondents and may also know their local environment best (Allen, 2006).

Therefore, based on the CDRI methodology, the AoRA was developed to understand the role or responsibility of different stakeholders (local government, communities, academia, private organisations and NGOs) in the implementation of defined action measures for Disaster Risk Reduction (DRR). Accordingly, the AoRA adopts the same five dimensions, but only 21 out of the 25 parameters which are contained in the CDRI (see Table 1). The reason for opting-out two parameters from the economic and natural dimension is as follows: firstly, aspects related to household assets and income are largely dependent on whether households have employment or not. This implies that action aiming at enhancing the economic resilience of communities requires an employed and skilful workforce. Once people have good jobs, income is obviously generated which ultimately leads to the acquisition of households assets. The strong relationship between income and household assets is presented in Table 9 of Chapter 5. Secondly, aspects related to the frequency and intensity of natural hazards are beyond the scope of how human actions can provide improvement or change.

Thus, 21 parameters are consisting each of them of three actions that have the potential to improve the climate disaster resilience of Chennai. The development of these DRR-related measures is not only based on the findings from the CDRI assessment in Chennai but also reflects practices which are regarded as positive in addressing deficiencies in the defined resilience dimensions and parameters (see Chapter 5). Furthermore, various on-site visits, interactions with local people, extensive literature review on lessons learned from past disaster events and guidance on DRR (Kyoto University (KU), 2010; United Nations International Strategy for Disaster Reduction (UNISDR), 2007) contributed in the process of identifying the action measures in the AoRA.

Hence, a total of 63 identified actions aim to reveal the importance of different actors in the implementation of selected action measures which

Table 1. Considered and Excluded Dimensions and Parameters for AoRA Based on the CDRI Methodology.

Dimensions	Physical	Social	Economic	Institutional	Natural
Parameters considered for AoRA	• Electricity • Water • Sanitation and solid waste disposal • Accessibility of roads • Housing and land-use	• Population health • Education and awareness • Social capital • Community • Preparedness during a disaster	• Employment • Finance and savings • Budget and subsidy	• Mainstreaming of DRR and CCA • Effectiveness of zone's crisis management framework • Knowledge dissemination and management • Institutional collaboration with other organisations and stakeholders • Good governance	• Ecosystem services • Land-use in natural terms • Environmental policies
Remaining parameters not considered in AoRA			• Income • Household assets		• Intensity/severity of natural hazards • Frequency of natural hazards

have the potential to enhance the disaster resilience of communities (wards). The practical approach of this assessment aims to find out to what extent different actions require a multi-stakeholder engagement or if a top-down, governmental-led planning is sufficient.

AoRA Process

The AoRA applied in Chennai sought to get the perceptions from the Councillors who are the members of the legislative body of the Chennai Corporation. Chennai Corporation means the municipality of Chennai. This legislative body consists of 155 members (Councillors) who are representing each of them one of the 155 wards of the city. The reason for choosing Councillors as a target group for the AoRA is due to their elected power to represent a majority of people within their constituencies. Thus, they are regarded as a type of community leader who is expected to represent the views of his/her community in a political environment. Furthermore, selecting Councillors as respondents to the AoRA aims to reveal how local decision-makers perceive the involvement of other stakeholders into the implementation of DRR-related action measures. Accordingly, three steps describe the structure and filling-up of the AoRA questionnaire:

1. For each of the 63 actions, the respondents (Councillors) need to decide on the level of implementation level, which is either 'not fully implemented' or 'fully implemented'. This step is needed to resolve doubts whether the proposed action measures are already in place based on the perceptions of the respondents.
2. The respondents (Councillors) have to decide about the level of responsibility of each identified stakeholder (local government, communities, academia, private organisations and NGOs) in the implementation of each of the proposed 63 actions. Thus, a range from 1 (very low), 2 (low), 3 (high) and 4 (very high) defines the scope of ranking the level of responsibility for each stakeholder. This step allows understanding how the respondents (Councillors) perceive which stakeholder should take the lead in the implementation of the proposed DRR-related actions. Accordingly, knowing about which stakeholders the local decision-makers (Councillors) regard as the most important in the implementation of the actions has the potential to smoothen the actual execution of the Climate Action Plan (CAP), presented in Chapter 8.

3. In a final step, the respondents (Councillors) have to prioritise which of the three actions for each of the 21 parameters should be implemented first. Thus, a simple 1, 2 and 3 ranking scale determines the prioritisation of actions, parameter-wise.

AoRA Results

The AoRA was conducted from September to November 2010 in collaboration with the Corporation of Chennai's Public Works Department which provided direct assistance in distributing the questionnaires to all the 155 Councillors with an attached letter from the Mayor requesting them to kindly respond to this call.

After the latest election in 2006, relevant for this study, the legislative body of the Chennai Corporation is represented by 56 women (36.1 per cent) and 99 men (63.9 per cent). However, it must be noted that the Chennai Corporation requires at least one third of the seats to be formally appointed by women, but in reality these women's husbands or other family members are often involved in the actual execution of the mandate in their constituencies. Thus, it is difficult to estimate to what extent unflavoured views of women are actually represented in the Council of the Chennai Corporation.

Implementation Level of Actions

The response rate of this survey was 83.2 per cent (total 129: women 49 [38 per cent], men 80 [62 per cent]) or 129 Councillors fully responded. Fig. 1 shows the implementation level for each of the 63 actions.

The results highlight that all actions are perceived to be not fully implemented according to at least 70 per cent of the Councillors. Thus, more than a qualified majority (66.67 per cent) of respondents believe that the proposed AoRA actions are not yet fully implemented in Chennai. This finding is crucial to elaborate on the second step of the AoRA as otherwise the remaining steps of the AoRA would have become obsolete. Although these results point out that taking action for building climate disaster resilience is highly needed in Chennai, it is surprising that all sectors are believed more or less equally to require improvement measures. This implies that Chennai is in need of a multi-dimensional strategy covering all aspects related to the functioning of the city. A potential strategy/policy/plan is discussed in the form of a CAP in Chapter 8.

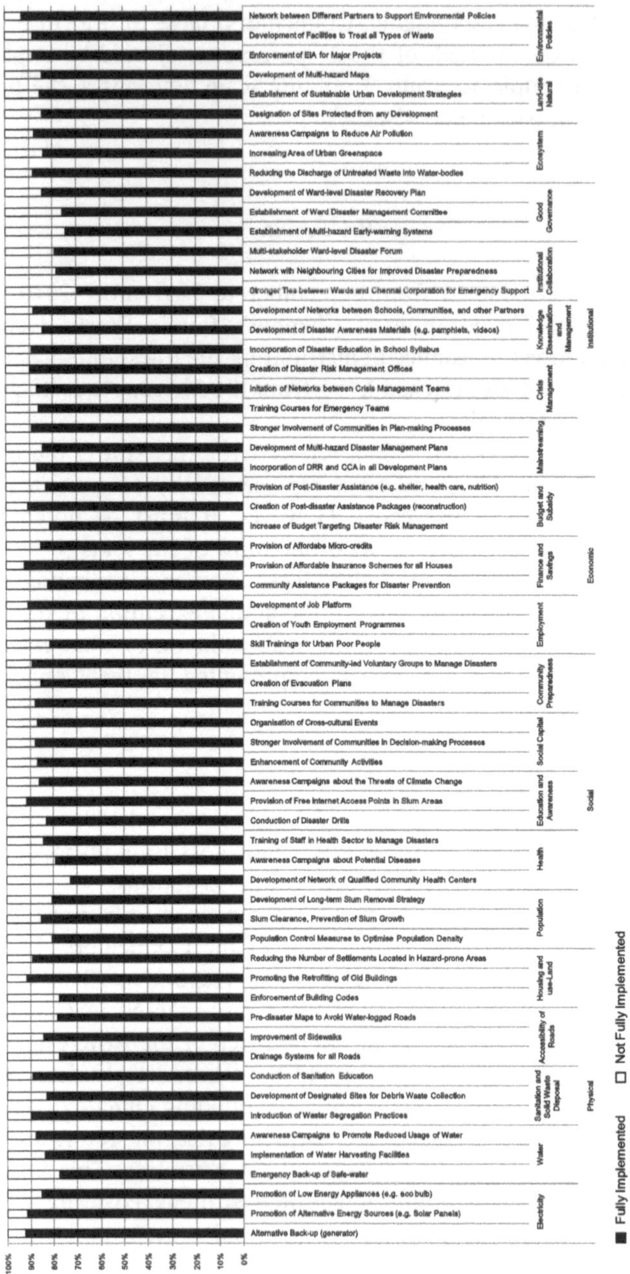

Fig. 1. Implementation Level of AoRA Actions.

Before, however, the role of different stakeholders involved in the implementation of the proposed AoRA actions aim to understand who should take the lead or highest responsibility in each of the 63 AoRA actions.

Responsibility Level of Stakeholders

In this second step, the responsibility level of the five identified stakeholders (local government, communities, academia, private organisations and NGOs) reveals who the Councillors perceive has highest responsibility in implementing each of the proposed 63 actions. Figs. 2–6 show the 'very high' responsibility scenario for each of the five dimensions. Hence, further graphical representations of the responsibility levels of 'high', 'low' and 'very low' are omitted. The reason to focus only on 'very high' level scenario is because of the initial objective that aims to reveal which stakeholders are most important in the implementation of defined DRR-related actions. Thus, other scenarios than the 'very high' one do not show new or compelling findings that would answer this objective.

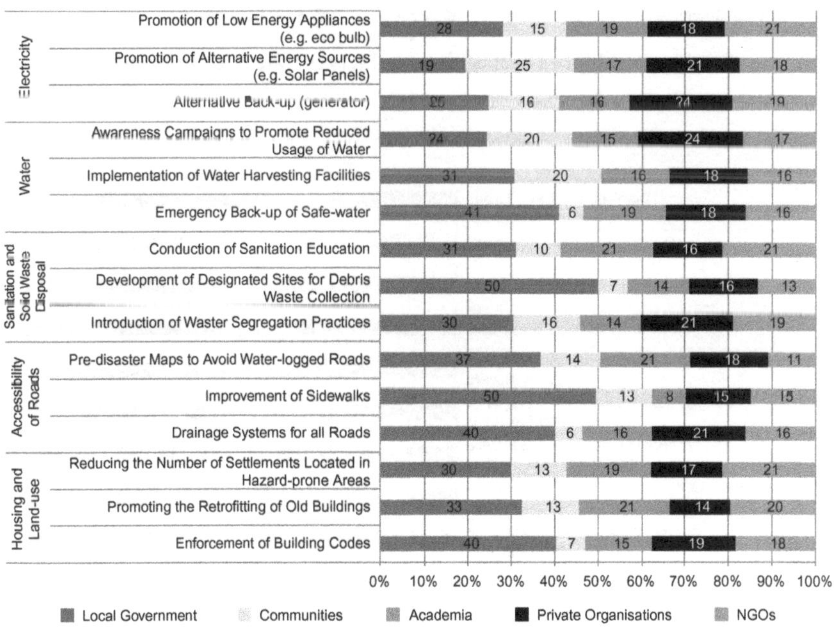

Fig. 2. Responsibility Level of Stakeholders for the Physical Dimension.

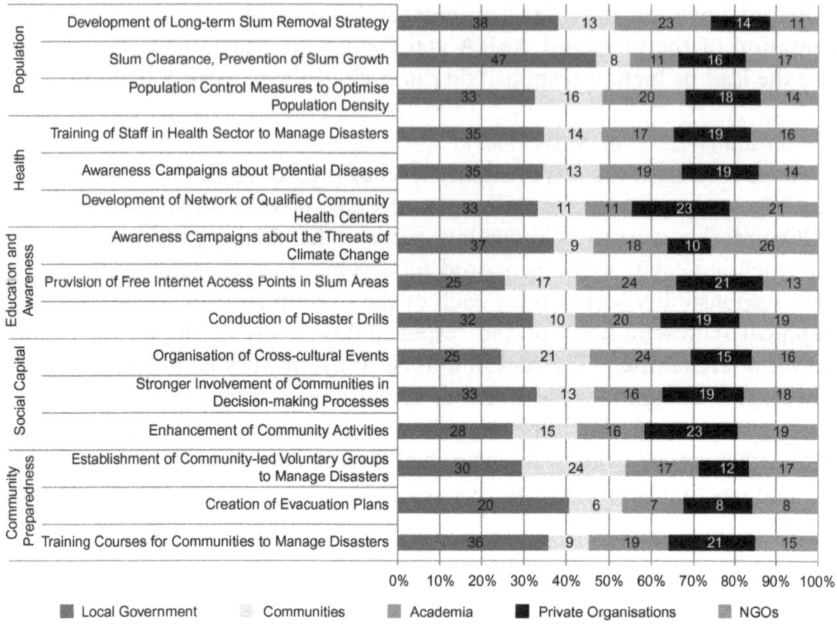

Fig. 3. Responsibility Level of Stakeholders for the Social Dimension.

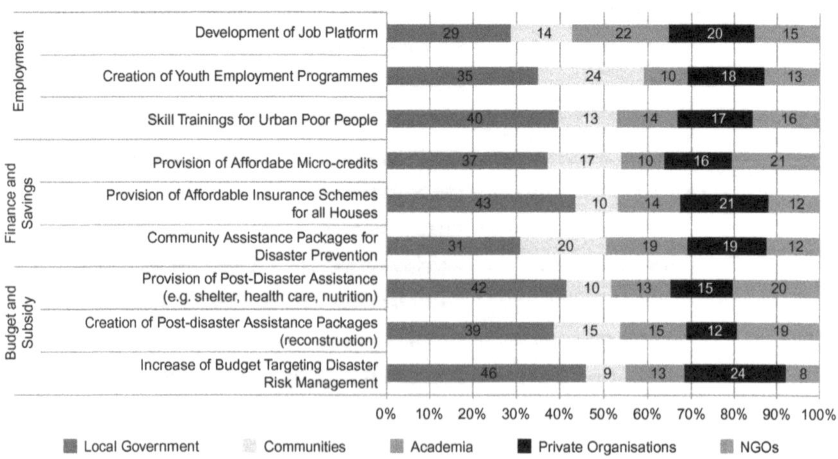

Fig. 4. Responsibility Level of Stakeholders for the Economic Dimension.

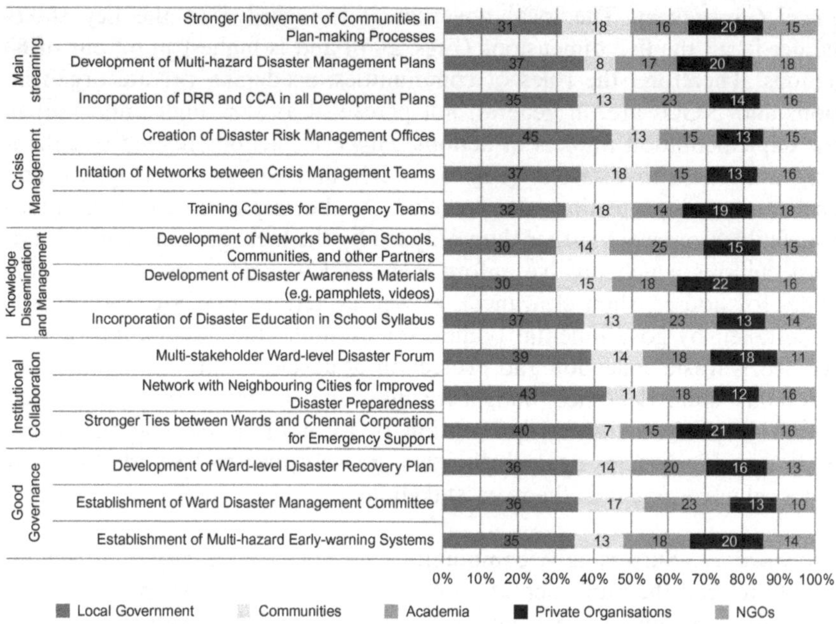

Fig. 5. Responsibility Level of Stakeholders for the Institutional Dimension.

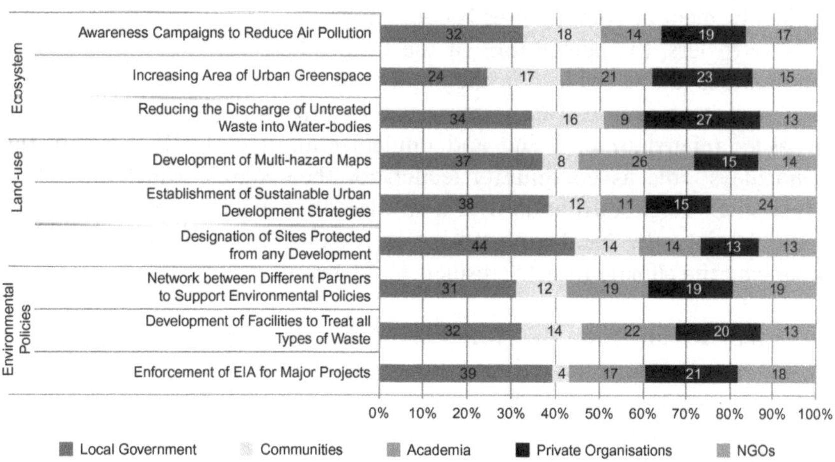

Fig. 6. Responsibility Level of Stakeholders for the Natural Dimension.

Local Government. The local government is regarded as the key stakeholder in all the five dimensions (Figs. 2–6) and is highest in 62 out of 63 actions. Therefore, the roles of communities, academia, private organisations and NGOs are, in general, not perceived as very high important in the implementation process of actions. However, actions related to enhancing awareness to disasters (Fig. 3), water and energy use (Fig. 2), and environmental protection are perceived to be implemented in a multi-stakeholder approach rather than only led by the local government. In contrast, actions which involve enforcement of laws and rules (e.g. building codes for houses, slum clearance), as well as actions that are traditionally undertaken by governmental bodies, for example, the designation of sites for debris waste collection and protection of development, improvement of sidewalks, slum clearance, budget for disaster risk management, as well as actions related to disaster prevention in form of developing plans and early-warning systems against disasters are perceived to be undertaken by the local government as the main stakeholder.

Communities. Surprisingly, communities are not given a very high responsibility to lead the implementation of actions, except in one (promotion of alternative energy sources) out of the 63 actions. Thus, communities are administered with little responsibility to carry out actions related to construction-related issues, for example, development of drainage systems or emergency back-up of safe water. However, in actions directed towards enhancing awareness to disaster and measures protecting the environment, the involvement of communities becomes more important.

Nevertheless, the perceptions of the Councillors which do not consider communities taking the lead in the implementation of more than one action to some extent demonstrate a conservative attitude from the respondents on issues related to planning and implementation of DRR. Despite the Councillors' role as community leaders of their wards, which are often rather small neighbourhoods with a size less than 1 km^2, their expectation is that mainly the local government is responsible to carry out actions enhancing the climate disaster resilience of Chennai.

Academia. Results from the AoRA show that the academia is regarded as an important partner in actions where scientific knowledge is needed, for instance, development of slum removal strategies, a job platform or hazard maps. Furthermore, the academia is recognised as a crucial partner in mainstreaming disaster education and DRR in development plans. However, the academia is not believed to be a leading stakeholder in any of the 63 actions. This may not be surprising as its role is clearly to provide information and knowledge to other stakeholders, such as the government when it comes to drafting disaster management plans. The academia's role is also an advisory one that supports, for example, the formation of a disaster management committee at the ward level (see Fig. 5).

Therefore, the academia is a crucial partner in the implementation of the proposed DRR-related actions.

Private Organisations. According to the AoRA, the involvement of private organisations is particularly important in actions belonging to the physical and natural dimension, for example, reducing the discharge of untreated water into waterbodies or the provision of alternative energy back-up systems. Furthermore, actions where the private sector can directly provide solutions, such as alternative back-up energy provisions, awareness campaigns to reduce the usage of water or establishment of free internet access points in slum areas, emphasise the expectation of the Councillors to give responsibilities to other stakeholders than only the local government. Councillors also perceive that the provision of funds for disaster risk management should receive contributions from the private sector.

Although private organisations are in none of the 63 actions regarded as the leading stakeholder, they must be regarded as crucial partner to be integrated into the implementation process. Thus, the findings also highlight that without the private sector, DRR cannot be implemented as actors such as private organisations are part of this process.

Non-governmental Organisations. NGOs are considered to take particular responsibility in actions related to developing sustainable urban development strategies, post-disaster assistance and awareness campaigns about threats of climate change. The findings further show that NGOs are regarded as crucial partner in provisions which are traditionally not or only partially delivered by the local government, such as affordable micro-credits, development of community health centres or post-disaster support.

Overall, NGOs are in none of the 63 actions the major implementer. This is to some extent surprising as NGOs would be ideal stakeholders to provide knowledge combined with small funds which could, for instance, support the raising of awareness to climate change and disaster risk. Little responsibility of NGOs is in actions aiming to increase the institutional resilience of Chennai. Nonetheless, like private organisations, NGOs must be considered as an important partner in DRR based on the findings from the AoRA.

Prioritisation of Actions

In the third step, the respondents (Councillors) were requested to prioritise the actions from the AoRA among each of the 21 parameters. Tables 2–6 show the prioritisation of the AoRA actions dimension-wise. The prioritisation was done based on the highest percentage of responses that were given to the three types of priorities: 1 (first), 2 (second) and 3 (third). In the analysis, a clear priority level (1, 2 or 3) was only determined if the particular action was at least 5 per cent points ahead of the next following

Table 2. Prioritisation of AoRA Actions for the Physical Dimension.

Relevance		Electricity	Priority
R	Action 1:	Alternative back-up at household level (e.g. *generator*)	NP
P, R	Action 2:	Promotion of alternative energy sources for owners of buildings/houses (e.g. *solar panels*)	NP
P	Action 3:	Promotion of low energy appliances (e.g. *eco bulb*)	NP
	Water		**Priority**
R	Action 1:	Establishment of *emergency back-up* of safe water supply for all citizens for a minimum period of 3 days	NP
P	Action 2:	Implementation of water harvesting facilities (e.g. *rooftop tanks*) for all buildings/infrastructures as a part of the building code	NP
P	Action 3:	Awareness programmes/campaigns to inform communities about the safe use of water (*reduce water usage*)	1
	Sanitation and Solid Waste		**Priority**
P, R	Action 1:	Introduction of *waste segregation practices* at household/community level	1
R	Action 2:	Development of *designated sites* for *debris waste collection* after a disaster	2
P, R	Action 3:	Conduction of *sanitation education* at the community level	3
	Accessibility of Roads		**Priority**
P, R	Action 1:	Develop all roads with adequate *drainage systems* which avoid water logging after intense rainfall events	NP
R	Action 2:	Widening and improvement of pedestrian walkways	2
R	Action 3:	Development of *pre-disaster maps* indicating roads vulnerable to water logging	NP
	Housing and Land-Use		**Priority**
P	Action 1:	Strict enforcement of *building permits* for all types of hazards for new buildings	1
P	Action 2:	Promoting the *upgrading/retrofitting* of older buildings	2
P, R	Action 3	Reducing the *number of settlements living on hazardous grounds* (relocation, provision of alternative housing)	3

Table 3. Prioritisation of AoRA actions for the Social Dimension.

Relevance	Population		Priority
P	Action 1:	Population control measures to optimise the *population density* at ward level	NP
P	Action 2:	Stronger *implementation of objectives outlined by the Tamil Nadu Slum Clearance Board*: slum clearance, prevention of slum growth, prevention of eviction of slum dwellers by private owners, etc.	2
P	Action 3:	Development of long-term *slum removal strategy* for the city of Chennai	NP
	Health		**Priority**
P, R	Action 1:	Development of *network of qualified community health centres* to serve underserved populations (e.g. urban poor)	1
P	Action 2:	Initiation of campaigns to raise awareness about *health concerns* (e.g. waterborne diseases) at household/ community level	2
R	Action 3:	*Trainings* for staff engaged in the health sector, particularly for situations of disaster	3
	Education and Awareness		**Priority**
R	Action 1:	Conduct *disaster drills* at ward level at least once a year	NP
P, R	Action 2:	Provision of free *internet access points* in slum areas	NP
P, R	Action 3:	Campaigns/awareness programmes *sensitising* the people about the threats of climate change	NP
	Social Capital		**Priority**
P, R	Action 1:	Enhancement of *community activities* through campaigns, projects and programmes (awareness-raising)	1
P	Action 2:	Stronger *involvement* of communities in *decision-making processes* (e.g. major projects, development plans, etc.) at ward level	2
P, R	Action 3:	Organisation of *cross-cultural events* linking different ethnic groups	3
	Community Preparedness during a Disaster		**Priority**
P, R	Action 1:	Provision of training courses and awareness programmes targeting communities for *efficient managing of disaster* situations	2
R	Action 2:	Creation of *evacuation plans* at the community level	1
R	Action 3:	Establishment of *community-led voluntary groups* at ward level for relief and recovery works (Community Emergency Response Teams)	3

percentage. If percentages fell within a 5 per cent gap to the next percentage, No Priority (NP) was determined. For example, the percentages for each of the three actions in the parameter water for the priority 1 were as follows: action 1, 28.68 per cent; action 2, 32.56 per cent and action

Table 4. Prioritisation of AoRA Actions for the Economic Dimension.

Relevance		Employment	Priority
P	Action 1:	Initiation of skill trainings for urban poor people	1
P	Action 2:	Creation of *youth employment programmes*	2
P	Action 3:	Develop network among different partners to establish a *job platform*	3
	Finance and Savings		**Priority**
P	Action 1:	Creation of comprehensive community assistance packages for *disaster prevention*	2
P, R	Action 2:	Provision of affordable *insurance schemes for all houses* to protect from damages of future disasters	1
P	Action 3:	Provision of *affordable micro-credits* for households	3
	Budget and Subsidy		**Priority**
P, R	Action 1:	Increase of *budget* targeting *disaster risk management* to 3% of overall Corporation budget by 2015	1
R	Action 2:	Creation of *post-disaster* assistance package for households whose houses require *rebuilding/reconstruction*	2
R	Action 3:	Provision of *post-disaster* assistance for severely affected households to *receive basic support* (*shelter*, *health care*, *nutrition*, etc.)	3

3, 38.76 per cent. As a result, action 3 was determined as the priority action among the three. Equally, the prioritisation for priorities 2 and 3 was made.

However, it must be acknowledged that this approach of analysis is semi-qualitative as it does not guarantee statistical significance ($p < 0.05$ significance level); therefore, it only shows some trends on how the Councillors perceive in which order the three proposed action measures in each of the 21 parameters should be implemented. In addition to the prioritisation of actions, the actions are characterised based on the time (relevance) when the implementation takes place and/or the proposed action should function. Thus, P stands for Preventive measure and R accounts for Responsive measure.

Analysing which action measures are prioritised so it becomes clear that DRR measures related to raising awareness (water use, community activities, skill training, training courses for emergency teams) and enforcement of laws (building code, EIA) are prioritised as primary actions. Furthermore, action measures targeting the building of financial capacities (insurance schemes for disaster protection, DRR financing) are also recognised as important measures to be implemented first.

Finally, prevention and the ability to respond to a disaster (preparedness) are key focus areas identified by the Councillors to be strengthened.

Table 5. Prioritisation of AoRA Actions for the Institutional Dimension.

Relevance	Mainstreaming of DRR and Climate Change Adaptation (CCA)		Priority
P, R	Action 1:	Incorporation of *DRR and CCA in all development plans,* especially in Master Plan	NP
P, R	Action 2:	Develop *disaster management plans* which address *all types of hazards* (multi-hazard risk assessments)	NP
P	Action 3:	Stronger *involvement of communities in plan-making processes*	NP
	Crisis Management		**Priority**
R	Action 1:	Increase of number of *training courses for emergency teams*	1
R	Action 2:	Initiation of *networks* between the different crisis management teams and other partners	2
P, R	Action 3:	Creation of disaster risk management offices	3
	Knowledge Dissemination and Management		**Priority**
P, R	Action 1:	Incorporation of *disaster education* in school syllabus	NP
P, R	Action 2:	Development of localised easy understanding disaster awareness materials (*pamphlets, videos,* etc.)	2
P	Action 3:	Development of *networks* between schools, communities and other partners to transfer knowledge about the threat of disaster and climate change	NP
	Institutional Collaboration		**Priority**
R	Action 1:	Development of stronger ties between the wards and Chennai Corporation for emergency support	1
P, R	Action 2:	Establishment of networks with neighbouring cities for improved disaster preparedness (e.g. allocation of resources)	2
P, R	Action 3:	Establishment of ward-level meetings/disaster forum with different partners (communities, NGOs, private sector, academia, etc.) for more efficient disaster management	3
	Good Governance		**Priority**
P, R	Action 1:	Establishment of efficient *multi-hazard early-warning systems* using radio, television and SMS technologies)	NP
R	Action 2:	Establishment of ward disaster management committee with participation of other partners to share and update emergency information during a disaster	2
R	Action 3:	Development of *ward-level disaster recovery plan*	NP

Summary of Results

The AoRA assessment revealed the following key findings:

1. No less than at least 70 per cent of all Councillors responded that the proposed 63 actions in the AoRA are not yet fully implemented (Fig. 1). This confirms that DRR is highly needed in Chennai.

Table 6. Prioritisation of AoRA Actions for the Natural Dimension.

Relevance		Ecosystem Services	Priority
P	Action 1:	Reducing the *discharge of untreated waste* into waterbodies (canals, rivers)	2
P	Action 2:	Increasing area of *urban greenspace* (trees, parks, etc.)	1
P	Action 3:	Campaigns/awareness programmes for communities to inform about the consequences of burning fossil fuels leading to *reduce air pollution*	3
	Land-Use		**Priority**
P	Action 1:	*Designation of sites protected* from any kind of development	NP
P	Action 2:	Establishment of sustainable urban development strategies for a balanced, mixed-use of available urban areas	2
P	Action 3:	Development of *multi-hazard maps* at ward level	NP
	Electricity		**Priority**
P	Action 1:	Strict enforcement of *Environmental Impact Assessments* for all major development projects	1
P	Action 2:	Development *facilities* to treat all types of waste (industrial, household, etc.)	2
P	Action 3:	Development of *network* between different partners to support environmental policies through awareness-raising and specific actions	3

2. The local government of Chennai has highest responsibility to implement 62 out of 63 actions (Figs. 2–6). In the only action (promotion of alternative energy sources) where the local government does not have the leading role, communities are regarded as the main stakeholder. This finding is very likely to represent the traditional perception of Councillors that the local government ought to have the leading in all issues related to DRR.
3. There is no action which is expected to be carried out by only one stakeholder (more than 50 per cent responsibility). Thus, there is a conviction among Councillors that a multi-stakeholder approach is required for all actions.
4. The prioritisation of actions emphasises that actions related to raising awareness (safe water use, community activities, skill training [job]), enforcement of laws (building codes, EIA) and the creation of improved financial capacities of households ought to be implemented first, or given highest priority.

Implications on Planning for Climate Disaster Resilient Communities in Chennai

According to the Councillors' perceptions, the local government is the key stakeholder in almost all of the 63 actions. This highlights a command-and-control attitude of local decision-makers where initiatives (actions) are launched and coordinated by an administrative authority (local government) instead of delegating certain actions to other stakeholders. Thus, the example from Chennai represents King's (2008) observations on how traditionally emergency management and urban planning is usually handled in developing countries.

Furthermore, these findings reflect the organisational structure of the local government and the planning agency in Chennai, whereby the latter is conducting little or no consultation with the public on how future planning scenarios of the city ought to look like. As a result, there are limited opportunities for communities to actively get involved into the decision-making process and implementation of actions which makes them highly dependent on other stakeholders, particularly the local government, to look after their needs. This lack of community involvement is visible in the AoRA where comparatively little responsibility is given to the communities to implement DRR as a leading stakeholder. Thus, their role is to carry out actions rather than being involved in the process of designing or shaping them. The consequences of this dependency is that communities may not receive sufficient support before, during and after disasters in case the local government fails to act appropriately; therefore, additional risk is created, making communities less resilient disaster events.

The limited power of communities in the political context of India's cities is emphasised by Baud and Nainan (2008) who assessed the opportunities of Advanced Locality Management (ALM) groups, established in Mumbai, as a successful approach that allows communities to become more active in activities related to DRR. These community groups, established in the 1990s mostly by middle class people, were supportive, for instance, in overcoming the devastating floods in Mumbai in 2005 as they provided relief and rescue operations.

Their role is to facilitate a government–community partnership and advocate the interests from their neighbourhoods in order to enhance the quality of life in these areas (Surjan & Shaw, 2009). Hence, Surjan and Shaw (2009) emphasise on the potential benefit of making use of the collective power of large number of people (e.g. community groups) in carrying out effectively specific DRR activities which require rather human resources than financial needs. This is a key reason why the AoRA focused more on soft adaptation measures for DRR instead of proposing hard measures. Among others, Allen (2006) and Van Aalst, Cannon, and Burton (2008) emphasise on the underestimated potential of communities to carry-out community-based activities which are close to many of the action measures defined in the AoRA and effective in reducing disaster risk.

The activities undertaken by the ALM groups are, for example, the collection of solid waste or cleaning of streets which reduce the risk of blocked drainage systems during intense rainfall events. However, Baud and Nainan (2008) conclude that little 'space' is given so far to these ALM groups to exert political power or get further involved into decision-making processes due to the hierarchical institutional set-up of Mumbai. Nonetheless, they gather people and carry-out community-based activities fostering DRR and thus, enhancing the resilience of communities to disasters. Moreover, this example has potential to be applied in other cities to increase the participation and consultation of local people into issues that take place at the very local level, in communities.

Although, the AoRA also shows that Councillors acknowledge that no action shall be implemented by solely by one stakeholder and thus, the involvement of other actors or partners is required in the implementation of DRR. For example, actions related to raising awareness to disasters and protecting the environment demonstrate clearly that the involvement communities, academia and NGOs are needed in the process of enhancing resilience in Chennai.

Limitations of AoRA Approach

The AoRA in Chennai focuses only on the perceptions of Councillors to determine the responsibility levels of the proposed actions; therefore, the findings are limited to views of just one group of stakeholders. Thus, the results do not reflect the views of the local government, academia, private organisations and NGOs, but only how local community leaders or political representatives perceive actions, related to building climate-related disaster resilience, to be implemented.

Therefore, depending on the political orientation of the Councillors, it cannot be ruled out that AoRA responses may become biased by their political opinions on the proposed action measures. However, choosing politicians as a target group for the AoRA allows understanding how local decision-makers perceive how their city can become resilient to climate-related disasters. Thus, they have considerable power to influence the actual implementation of actions at the city level, and therefore their views are expected to have significant weight for building a climate-related disaster resilient city. Nevertheless, it must be acknowledged that the AoRA could be further conducted with other stakeholders who would provide a greater range of different perceptions on this issue.

The AoRA is a semi-qualitative approach to enhance actions at the city level. This may not be directly applicable in other cities if a CDRI assessment is not undertaken beforehand, as the selection of actions depends on the needs of improvement which are identified through the CDRI. Furthermore, the participatory process of involving local decision-makers and stakeholders in selecting and identifying the priority actions is regarded as crucial in the process of making plans and strategies for the future development of the city.

The identification of action measures in the AoRA is based on the interpretation of existing literature which gives orientation on how to enhance the climate disaster resilience of communities. Therefore, actions are tailored to address climate-related disaster and do not focus on enhancing the resilience to other existing types of natural hazards. Finally, the AoRA may not include each and every required action that would be needed to cover all aspects of building climate-related disaster resilience.

Summary of AoRA

To summarise, the AoRA has the potential to precisely determine the roles of different actors in implementing specific actions. Furthermore, it can support the implementation of the Basic Services to the Urban Poor (BSUP), launched in 2009, in Chennai. Although the Corporation of Chennai is committed and eager to actively address the problematic issue in the many slums of the city, through the BSUP, without the support of community groups and NGOs, it is difficult to effectively improve the conditions in the urban poor areas. Another example where the AoRA can support the city planning is through the promotion of rain-water harvesting facilities. Although the Chennai Metropolitan Water Supply and Sewerage Board (CMWSSB) promote such installations, little has been undertaken to increase the number of rain-water harvesting facilities in Chennai so far. Thus, the AoRA may give emphasis on how Councillors, in this case, perceive who and what level of responsibility different stakeholders have in facilitating such installations, which in this case requires clearly a multi-stakeholder approach (see Fig. 2).

Finally, through the AoRA an attempt was made to understand the responsibilities of five key stakeholders in building resilience to climate-related disasters in place-bound communities (wards) in Chennai. The results from the proposed resilience enhancing actions show that the local government is seen as the most important actor to implement actions which are related to disaster prevention and risk reduction. However, communities, academia, private organisations and NGOs are depending on the type of action increasingly encouraged to take responsibility in form of engaging into partnerships. Moreover, the prioritisation of actions pointed out that a combination of actions related to raising awareness about disasters and the provision of financial capacities to protect from such events are prioritised.

REFERENCES

Allen, K. M. (2006). Community-based disaster preparedness and climate adaptation: Local capacity-building in the Philippines. *Disasters*, *30*(1), 81–101.

Baud, I., & Nainan, N. (2008). 'Negotiated spaces' for representation in Mumbai: Ward committees, advanced locality management and the politics of middle-class activism. *Environment and Urbanization*, *20*(2), 483–499.

King, D. (2008). Reducing hazard vulnerability through local government engagement and action. *Natural Hazards*, *47*, 497–508.

Kyoto University (KU). (2010). *A guide for implementing the Hyogo framework for action by local stakeholders*. United Nations, Geneva: Consultation version.

Surjan, A., & Shaw, R. (2009). Enhancing disaster resilience through local environment management. *Disaster Prevention and Management*, *18*(4), 418–433.

United Nations International Strategy for Disaster Reduction (UNISDR). (2007). *Words into action: A guide for implementing the Hyogo framework*. Geneva: United Nations.

Van Aalst, M. K., Cannon, T., & Burton, I. (2008). Community level adaptation to climate change: The potential role of participatory community risk assessment. *Global Environmental Change*, *18*, 165–179.

CHAPTER 7

PERCEPTIONS OF HOUSEHOLDS ABOUT ENHANCING THE CLIMATE DISASTER RESILIENCE OF COMMUNITIES IN CHENNAI, INDIA

INTRODUCTION

After the results from the AoRA, presented in Chapter 6, emphasised on the limited involvement of communities based on Councillors' perceptions, further investigations at the micro-level (households) of communities seek to reveal the challenges and potential to attract more community participation into DRR. Additionally, the two case studies from selected wards in Chennai serve to provide a more detailed understanding about how communities in Chennai function and what their perceptions are towards climate-related disasters. Moreover, the conducted household survey in Ward 79 (Adikesvapuram) and Ward 131 (South Kodambakkam) aims to provide further insights on the preparedness and capacity of individuals from these two neighbourhoods to respond to climate-related disasters.

Finally, the following research questions summarise the challenges faced in enhancing the climate disaster resilience of Chennai and ought to be answered in the following sections

- Are households located in the vicinity of natural hazards (rivers, canals) less resilient?
- Do affected households learn from climate-related disasters and enhance their resilience?
- How can community resilience to climate-related disasters be enhanced?

COMMUNITY DISASTER RESILIENCE

Although community resilience has already been discussed in Chapter 3, further elaborations on this issue are needed to fully understand the implications on enhancing a community's resilience through urban planning decisions.

Based on the life-cycle model of Cutter et al. (2008), community resilience to disasters encompasses various time periods in which different abilities of communities determine their actual resilience to disasters. In other words, it is argued that the abilities of communities to exert adaptive capacity depend on their available coping capacity during time of disturbance (disaster).

In this context (disasters), the term adaptive capacity is described by Gallopin (2006) as the ability of socio-ecological systems, including communities, to learn and improve their capacity either through reaction due to a disturbance or in a proactive manner where a future stress or change is anticipated before it occurs. Accordingly, higher experience of disasters (flood-related) enhances the preparedness and learning capacity of people.

Community Disaster Resilience Model

The previous notions on the coping and adaptive capacities are crucial in the understanding about what determines the resilience of communities. Therefore, based on the literature review, a community disaster resilience model (Fig. 1) aims to reveal and highlight the various theoretical components which determine the ability of an individual person to anticipate, absorb, manage and recover from a disaster, as emphasised by Twigg (2007).

Accordingly, this community disaster resilience model (Fig. 1) has been developed to highlight the capacity or condition of a community during different time phases before, during and after a disaster. Thereby, the existence of adaptive capacity is crucial, as mentioned before, to enhance a community's ability (resilience) to confront and manage future disasters. Hence, three potential scenarios can evolve due to a disaster:

- *A*: The community's coping capacity is not sufficient to absorb the hazard. As a result, the community suffers from damages and losses and its condition becomes weak. During the recovery phase a community can get back to its pre-disaster condition (middle way), but also either not

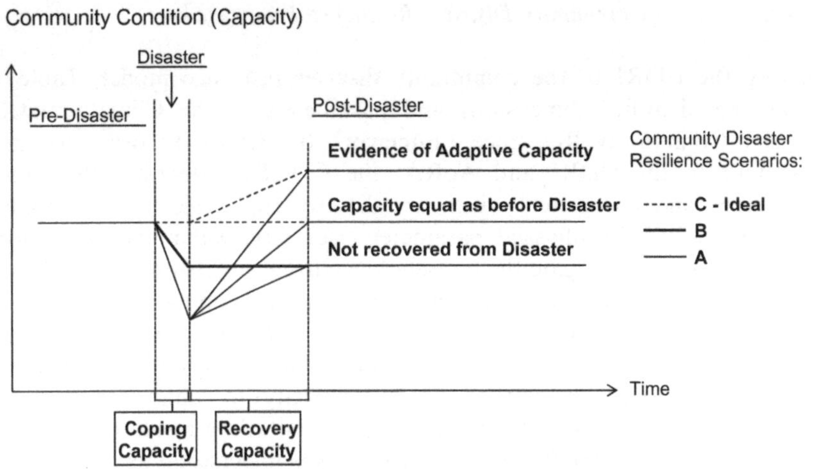

Fig. 1. Community Disaster Resilience Model (Graphical).

fully recover or learn and adapt (ideal) from the disaster and enhance its resilience (capacity).

- *B*: The community's coping capacity is not sufficient to absorb a hazard. As a result, the community suffers from damages and losses and its condition becomes again weak. However, unlike scenario A, the community cannot exercise recovery action and neither adaptive capacity and thus, remains weak.
- *C* (*ideal scenario*): The community's coping capacity is sufficient enough to absorb a natural hazard (rainfall, storm, earthquake, etc.) and therefore, no damages or losses are recorded. Although the community may not have been affected by damages and losses, it may still exercise a certain amount of adaptation (adaptive capacity) to strengthen its coping capacity for potential disaster events in the future; otherwise, the capacity remains the same as before the natural hazard occurred.

The graphical representation of community disaster resilience emphasises that the levels of a community's coping and adaptive capacity are crucial in absorbing, managing and bouncing back from a disaster event. Particularly, the ability or capacity of a community to adapt to disasters is crucial in enhancing their resilience.

This theoretical excursion clears the path to link the concept of resilience applied in the CDRI of larger urban units (cities, zones) to neighbourhoods and thus, allows understanding resilience of individuals (households).

Community Disaster Resilience Framework

Linking the CDRI to the community disaster resilience model, Table 1 shows the identified dimensions and parameters of the Climate-related Disaster Community Resilience Framework (CDCRF). In order to stay consistent to the CDRI and AoRA, the CDCRF integrates the same dimensions and parameters which are directly applicable to households. For example, in the physical parameter electricity, water and sanitation and solid waste, investigate the household's current provision of these basic services. This shall serve to understand their physical coping capacity.

Since the community disaster resilience has a second component, recovery capacity, the CDCRF aims to understand quantitatively whether affected households take action (adaptive capacity) to enhance their resilience as a reaction to experienced disaster(s) or not. As a result, community disaster resilience in the following case studies is understood as a combination of coping and recovery (adaptive) capacity.

The theoretical linkage between the community disaster resilience model and the CDCRF is summarised as follows: to achieve scenario C (Fig. 1) from the community disaster resilience model, the parameters from the three dimensions of the CDCRF need to be met at a high level, for example, full knowledge among households on impacts of climate change in the parameter of education and awareness (social dimension). However, in reality, this may not be the case for all households and therefore households are asked whether they take action (adaptive capacity) to enhance their knowledge to be better prepared for future climate-related disasters. This issue of taking action has particular importance for those households who were affected by climate-related disasters as their resilience (coping capacity) was not sufficient to fully absorb the disaster. Thus, it is assumed that affected households have a particular incentive to take action in order to avoid future impacts from natural hazards. In the case that disaster affected households take action, they would follow scenario A, according

Table 1. Dimensions and Parameters of the CDCRF.

Dimensions	Physical	Social	Economic
Parameters (P)	Electricity	Health	Income and employment situation
	Water	Education and awareness	Household assets
	Sanitation and solid waste	Social capital and preparedness	Finance and saving

to the community disaster resilience framework (Fig. 1), which is also a desired (ideal) scenario if they demonstrate adaptive capacity which would increase their resilience beyond the level they had before the disaster event.

Limitations of CDCRF Approach

Similar to the CDRI and AoRA, the CDCRF has certain limitations which need to be mentioned. Firstly, the number of parameters in the physical, social and economic dimension may be further increased, however, to understand the behaviour and ability of people to respond to disasters the selected parameters are expected to be sufficient Secondly, the CDCRF assesses the resilience of communities during non-disaster times, implicating that attributes of community resilience, to absorb, manage and bounce back following a disaster, are analysed based on their current capacity. Thirdly, the presented resilience framework neglects potential involvement from other stakeholders (e.g. local government) to support communities during a disaster. Instead, the focus is put entirely on how individual households are likely to manage climate-related disasters themselves and whether they have learned from previous experiences and increased their coping capacity through adaptation processes.

CDCRF IN TWO COMMUNITIES OF CHENNAI

Characteristics of Study Locations

Based on the CDRI assessment, two comparable neighbourhoods were selected to undertake a household survey related to the CDCRF. Since the results from the CDRI revealed that Zone VI has the lowest overall resilience and Zone IX highest, one ward from each of the two zones was selected to understand potential differences regarding their disaster resilience.

As a result, Ward 79 (Adikesvapuram) and Ward 131 (South Kodambakkam), which are both located along major rivers of Chennai and are predominantly residential, were selected as the two study sites for the household survey (Fig. 2).

The key characteristics of the two selected sites (Fig. 3) are as follows:

- Ward 79 belongs to the historical part of Chennai (built during the 19th century) unlike Ward 131 which is located in new developed area that became from the 1970s onwards.

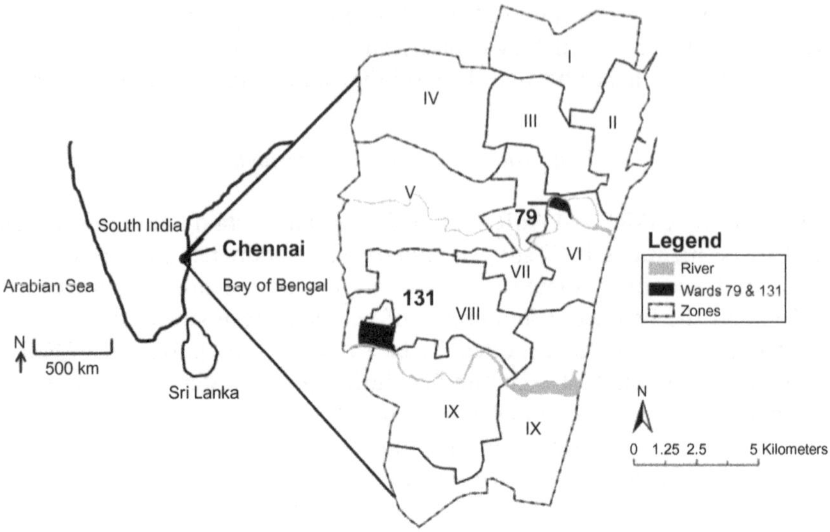

Fig. 2. Location of Selected Wards 79 and 131.

Fig. 3. Comparative Visual Impressions on Study Locations. *Source*: Pictures by Author.

- The area sizes of Ward 79 and 131 are 0.45 km^2, respectively 1.26 km^2, in which the former had 18,190 inhabitants and the latter had 58,400 in 2011. Therefore, the population densities in both wards go beyond 40,000 people per square kilometre. These population densities are considerably high compared to Chennai's average of 26,762 (Chennai Metropolitan Development Authority (CMDA), 2008).
- The disaster profiles of both wards are similar not only because they are located at a river, but also because of their close proximity to each other which makes them equally prone to impacts of cyclones in form of intense rainfall events leading to potential flooding throughout Chennai during the post-monsoon period from October to December (Drescher et al., 2007; Revi, 2008).
- Although Chennai suffered from past cyclone-related floods in 2005, 2008 and 2010, residents from Ward 131 claim that major floods occurred in the years 1976, 1983, 2006, 2008 and 2009, mostly after dams of lakes in the upstream of the Adyar River were opened-up to avoid overflowing after intense rainfalls. Thus, the flooding was to some extent man-made.
- In addition to big events, yearly small-scale flooding lead to temporary water-logging in both wards and perceived by some residents as flood events as these type of events have as well potential to cause damages to people and houses.

The above characteristics of the two selected study sites highlight that both are residential but were developed during different periods and are likely to be challenged differently by impacts of urbanisation.

Design of Data Collection

A household survey was undertaken during May to June 2011 through a combination of systematic and random sampling techniques. Since the availability of secondary data, such as numbers of households or numbers of houses in each street, do not exist in both wards which would support a pre-planned approach in conducting a household survey, an alternative way had to be chosen.

Therefore, in order to get a spatially equal representation of household responses in each ward, but still conform to commonly accepted survey techniques, every third house of all streets in both wards was selected as a sample. However, residents from slum and urban poor areas were excluded from this study. Households were approached during all days of a week usually between 4 and 7 pm to conduct face-to-face interviews by filling up a standardised questionnaire which is based on the CDCRF.

The total sample size in Ward 79 was 336 (8.4 per cent of all households [average household size was derived from survey]) and 1,112 for Ward 131 (8.7 per cent of all households) and the response rate was 58.0 per cent in Ward 79 and 59.4 per cent in Ward 131.

Results of Household Survey in Wards 79 and 131

In the following sections, the results are compared between households whose houses were damaged by a climate-related disaster in the past and those whose houses have not been affected. The splitting into these two groups is a viable approach to respond to the research questions set in the introduction of this chapter to understand where the disaster-prone areas are and also whether households have learned from such experiences.

Although it may be criticised that determining disaster experience based on whether a household's house was damaged or not does not include all households who have experienced a disaster. However, in order to achieve the objective of spatial analysis about where households are located in disaster-prone areas, the determination of disaster experience is based on damages on a household's home. This determination is supported by the fact that 78.6 per cent of households in Ward 79 and 83.0 per cent in Ward 131 claim to have had a past climate-related disaster experience and also have had their houses damaged. As a result, there is a strong correlation between households whose houses were damaged and households who have experienced (e.g. injuries, economic loss, etc.) a climate-related disaster in the past. The statistical correlation coefficient for Ward 79 is $r = 0.82$ and $r = 0.83$ for Ward 131.

Spatial Analysis of Disaster-Prone Areas

Results from the household surveys in Wards 79 and 131 point out that flood was in more than 92 per cent of all responses the climate-related disaster type that affected households. Thus, other climate-related natural hazards such as storms, droughts or heat waves are hardly experienced by residents in the two selected neighbourhoods. Secondly, 28.7 per cent in Ward 79 and 43.6 per cent of all households in Ward 131 claim to have experienced a climate-related disaster, mostly flood, in the past, but have not necessarily been affected. However, as mentioned before, the relevant figures in the analysis of the two wards regarding households who were affected (house damaged) show that *24 per cent in Ward 79* and *37 per cent in Ward 131* have experienced damages on their houses due to mostly floods.

Fig. 4 shows the locations of houses (in red) to which households claim to have suffered from climate-related disaster damages in the past in Ward 79, the equivalent for Ward 131 is shown in Fig. 5. Both maps (Figs. 4 and 5) have been drawn by using ArcGIS.

Interestingly, the distribution of damaged houses is particularly profound in areas located in the vicinity of waterbodies (river, canal). Hence, it can be concluded that living locations near rivers and canals are more disaster-prone than those further away. As Fig. 5 shows large slum settlements border directly the river; therefore, it must be expected that people living in those areas are even more affected by flooding events than those houses which are a couple of ten metres away from the river.

This finding answers the first out of the three research question and provides a starting position to investigate whether households who were affected by damaged houses due to climate-related disasters in the past have learned from them and became better prepared. Additionally, the

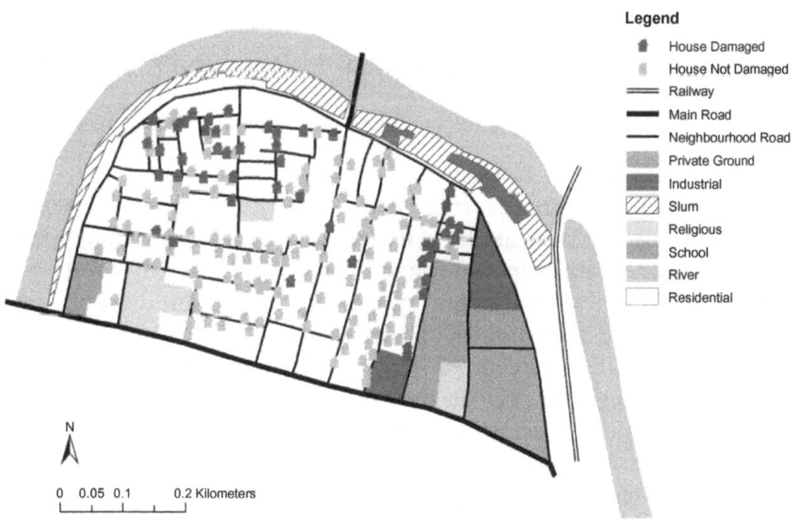

Fig. 4. Location of Damaged and Not Damaged Houses Due to Climate-Related Disasters in the Past in Ward 79.

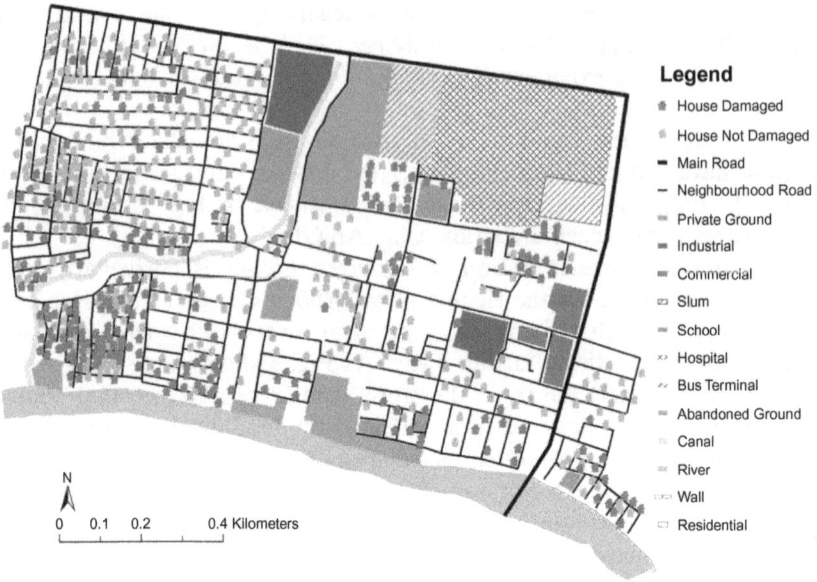

Fig. 5. Location of Damaged and Not Damaged Houses Due to Climate-Related Disasters in the Past in Ward 131.

maps show the population density of the two selected wards, as all samples have been mapped-out.

Overall Characteristics of Households

Overall, basic characteristics (sex, marital status, etc.) of households' responses (Table 2) show similar patterns between the two communities, but significantly (statistical significance based on two-sided *t*-test) more people who always lived (no migration) in Ward 131 claim disaster experiences that damaged their houses. The reason for this is likely to be two-fold: on one hand, they have lived in average longer in the neighbourhood than those who migrated, but on the other may not be willing to relocate their living location. Interestingly, in both wards, residents who have experienced a disaster (flood) have significantly lower English abilities (speaking and reading) compared to those who claim no damage to their houses due to climate-related disasters (floods). This is an indication that less privileged residents in terms of school education are also less resilient.

Table 2. Overall Characteristics of Households (HH) in Wards 79 and 131 (Bold Values Are Marking Statistical Significance to Be Different; $p < 0.05$).

Key Characteristics		Ward 79		Ward 131	
		House not damaged ($n=149$) in %	House damaged ($n=46$) in %	House not damaged ($n=416$) in %	House damaged ($n=245$) in %
Sex	Male	44.3	28.3	36.3	29.8
	Female	55.7	71.7	63.7	70.2
Age	15–24	12.1	10.9	**20.7**	**11.0**
	25–64	72.5	82.6	67.3	67.3
	More than 65	15.4	6.5	**12.0**	**21.6**
Marital Status	Single	16.1	13.0	**17.3**	**11.0**
	Married	83.9	87.0	82.7	89.0
Household Size	Average # (not in %)	4.6	4.9	4.4	4.3
Migration	From Tamil Nadu State, but outside of Chennai	7.4	8.7	**34.4**	**24.9**
	From outside of Tamil Nadu State	4.0	2.2	5.8	2.9
	Within Chennai	42.3	39.1	37.0	31.0
	No migration	46.3	50.0	**22.8**	**41.2**
Mother Tongue	Tamil	80.5	93.5	89.4	91.0
Languages (Speaking)	Tamil	98.7	95.7	99.3	99.2
	English	**53.0**	**13.0**	**58.9**	**40.8**
Languages (Reading)	Tamil	98.7	97.8	98.6	98.4
	English	**54.4**	**13.0**	**59.9**	**41.2**

Results on Physical Resilience

Results in the physical resilience (Table 3) reveal that households who experienced damaged houses due to climate-related disasters suffer more from power cuts (in both communities) than others and are in particular less satisfied with the electricity provision in Ward 79. Other findings emphasise on the aspect that residents with disaster experiences due to damaged houses are possibly less privileged, for example, they have in both communities significantly lower provision of water storage tanks. Moreover, the garbage for residents in Ward 131 is collected at fewer households and also less frequently. In terms of adaptive capacity, residents who have experienced damages to their houses due to disasters are not taking more action compared to those who claim no disaster damages to their houses. This is likely due to the relatively high provision of basic services, meaning electricity, water, sanitation and solid waste management services are supplied at a high level.

To summarise briefly, households with disaster experiences are not more resilient in relation to physical aspects compared to others who do not claim damages to their houses due to a climate-related disasters.

Results on Social Resilience

Results from the household surveys (Table 4) in the two selected communities in Chennai highlight that residents with disaster experience (house damaged) are significantly more affected by waterborne and vector-borne diseases compared to those who have not been impacted by damages on their houses due to climate-related disasters.

In contrast to the lower English ability (reflects their education level) of disaster affected (damage on house) households (Table 2), their perception about knowing the impacts of climate change and also their awareness on how to respond to climate-related disasters is not significantly different compared to households who claim no damages to their houses due to climate-related disasters. Regarding aspects of social capital and disaster preparedness, households in Ward 79 with disaster experience (house damaged) are on one hand more actively involved in Community-based Organisations (CBOs) than residents who have not experienced climate-related disaster damages to their houses on the other, they are significantly less prepared to face such disasters as their possession of basic emergency equipments (medical kit, flash light, food, etc.) is available among less households. Looking at aspects of adaptive capacity, in none of the social parameters residents with disaster experience (house damaged), in both communities, show efforts to take action to increase their resilience.

Table 3. Physical Resilience of Households (HH) in Wards 79 and 131 (Bold Values Are Marking Statistical Significance to Be Different; $p < 0.05$).

P	Key characteristics		Ward 79		Ward 131	
			House not damaged ($n=149$) in %	House damaged ($n=46$) in %	House not damaged ($n=416$) in %	House damaged ($n=245$) in %
Electricity	**HH supplied with electricity?**	Yes	100	100	99.5	99.2
	Power cuts:	More than once a day	**36.2**	**63.0**	**34.5**	**49.8**
		Once a day	**63.1**	**37.0**	**65.5**	**49.0**
		Less than once a day	0.7	0.0	0.0	0.0
	HH satisfied with electricity provision?	No	**36.9**	**69.1**	37.3	42.0
	If no, HH take action?	No action	**(n=55) 60.0**	**(n=18) 88.9**	($n=155$) 69.0	($n=103$) 67.0
Water	**HH supplied with piped water**	No	**6.7**	**21.7**	6.0	9.4
	If yes, HH has water storage tank?	Yes	**(n=139) 84.9**	**(n=36) 58.3**	**(n=391) 80.6**	**(n=222) 69.4**
	HH satisfied with water provision?	No	5.4	6.5	7.5	13.9
	If no, HH take action?	No action	($n=8$) 87.5	($n=3$) 66.7	($n=31$) 38.7	($n=34$) 35.3
Sanitation and solid waste	**HH connected to a drainage system?**	Closed drainage	98.0	95.7	99.8	98.8
	HH satisfied with sanitary provision?	No	**13.4**	**34.8**	**5.0**	**10.6**
	If no, HH take action?	No action	($n=20$) 55.0	($n=16$) 56.3	($n=21$) 95.2	($n=26$) 84.6
	HH's garbage collected?	No	5.4	2.2	**2.4**	**13.5**
	If yes, how often?	Once every day	86.5	86.7	**84.0**	**75.5**
		Once every three days	9.2	4.4	**14.0**	**22.2**
		Less than once every three days	4.3	8.9	2.0	2.3
	HH satisfied with waste management?	No	15.4	8.7	**11.3**	**23.3**
	If no, HH take action?	No action	($n=23$) 91.3	($n=4$) 100.0	($n=47$) 76.6	($n=57$) 61.4

Table 4. Social Resilience of Households (HH) in Wards 79 and 131 (Bold Values Are Marking Statistical Significance to Be Different; $p < 0.05$).

P	Key characteristics		Ward 79		Ward 131	
			House not damaged (n = 149) in %	House damaged (*n* = 46) in %	House not damaged (*n* = 416) in %	House damaged (*n* = 245) in %
Health	**Affected by WD in the past 5 years?**	Yes	**10.1**	**45.7**	**13.9**	**33.9**
	If yes, take action?	No action	(*n* = 15) 13.3	(*n* = 21) 4.8	(*n* = 58) 6.9	(*n* = 83) 2.4
	Affected by VD in the past 5 years?	Yes	**21.5**	**54.3**	**24.5**	**38.8**
	If yes, take action?	No action	(*n* = 32) 3.1	(*n* = 25) 0.0	(*n* = 102) 8.8	(*n* = 95) 9.5
Awareness	**Knowledgeable about impacts of climate change?**	No	55.7	69.6	52.6	55.1
	If no, take action?	No action	(*n* = 83) 77.1	(*n* = 32) 90.6	(*n* = 219) 72.1	(*n* = 135) 76.3
	Knowledgeable on how to respond to a CRD?	Not fully	75.8	73.9	72.8	79.6
	If not fully, take action?	No action	**(*n*=113) 65.5**	**(*n*=34) 88.2**	(*n* = 303) 54.5	(*n* = 195) 54.9
Social capital and preparedness	**Participation in CBOs?**	Yes	**8.7**	**23.9**	30.5	33.9
	HH satisfied with the services provided by Municipality?	No	26.2	23.9	**12.3**	**24.5**
	If no, take action?	No action	(*n* = 39) 94.9	(*n* = 11) 100.0	(*n* = 51) 86.3	(*n* = 60) 83.3
	HH prepared for a CRD (disaster supply kit)?	Yes	**43.6**	**13.0**	36.1	32.7
	HH capable to provide voluntary support during a CRD?	Yes	49.7	45.7	63.2	58.8

To conclude briefly, households with disaster experience (house damaged) are less resilient regarding health aspects, but are equally resilient in issues related to parameters on awareness and social capital and preparedness. Unfortunately, there are no indications that households are learning or taking action due to a disaster experience to increase their social resilience.

Results on Economic Resilience

The economic resilience shows patterns (Table 5) that household with disaster experience (house damaged) derive more income from the informal sector, which means during disaster times the income generation may be less secured compared to employment in the formal sector. Accordingly, it may not surprise that less people are satisfied with their employment situation which is likely to provide irregular income in the informal sector. However, the adaptive capacity of these concerned people is limited as significantly less action is taken by those living in Ward 131 to apply for jobs or search for new income sources. Regarding household assets, residents with disaster experience (house damaged) have less access to computer with internet compared to those who have not been affected by damages on their houses due to climate-related disasters. This exemplifies again that residents with disaster experience (house damaged) are less privileged to afford services, such as internet at house – lower coping capacity. Aspects of finance and savings point out that the ability of residents with disaster experience (house damaged) to save money or insure their house to climate-related disaster is either equally low or again lower in Ward 131 (adaptive capacity) compared to non-affected (house damaged) households.

To summarise, households with disaster experience (house damaged) are less resilient in all the three parameters of economic resilience compared to residents who have not been affected by climate-related disasters (house damaged). Furthermore, households do not learn or take action due to past disaster experiences.

Summary of Results

In an attempt to quantitatively assess the resilience of communities to climate-related disasters in two neighbourhoods with similar characteristics (exposure to natural hazards and land-use) of Chennai, the ability of households to learn from disaster experiences is measured as low. Results from the household surveys In Wards 79 and 131 show that households who experienced disasters (house damaged) are less privileged, for example, in terms of level of education, access to internet from home or having jobs in the formal sector (to get regular income) compared to those who have not

Table 5. Economic Resilience of Households (HH) in Wards 79 and 131 (Bold Values Are Marking Statistical Significance to Be Different; $p < 0.05$).

P	Key characteristics		Ward 79		Ward 131	
			House not damaged ($n = 149$) in %	House damaged ($n = 46$) in %	House not damaged ($n = 416$) in %	House damaged ($n = 245$) in %
Income and employment	**HH gets income from formal sector?**	Not fully	**63.1**	**89.1**	57.9	56.3
	If not fully, from where?	Informal sector	($n = 94$) 85.1	($n = 41$) 92.7	**(n=241) 71.4**	**(n=138) 87.0**
		Formal and informal sector	($n = 94$) 14.9	($n = 41$) 7.3	**(n=241) 28.6**	**(n = 138) 13.0**
	Satisfied with employment?	No	**22.1**	**41.3**	**14.2**	**26.5**
	If no, take action?	No action	($n = 33$) 51.5	($n = 19$) 68.4	**(n=59) 11.5**	**(n=75) 78.5**
Household assets	**HH equipped with HH assets?**	Yes	100.0	97.8	99.8	98.4
	If yes, with which devices?	Radio	**36.2**	**(n=45) 17.8**	($n = 415$) 27.0	($n = 241$) 20.3
		TV	100.0	($n = 45$) 100.0	($n = 415$) 99.3	($n = 241$) 99.6
		Computer internet	**28.2**	**(n=45) 4.4**	**(n=415) 35.7**	**(n=241) 26.1**
	Key HH assets protected from CRD?	Yes	69.1	67.4	87.7	87.8
Finance and savings	**House insured against CRD?**	No	97.3	100.0	95.9	93.9
	If no, take action?	No action	($n = 145$) 93.8	97.8	($n = 399$) 94.9	($n = 230$) 89.1
	HH saves money for future CRD?	No	83.9	91.3	**82.0**	**89.8**
	If no, get support?	Get no support	($n = 125$) 51.2	($n = 42$) 42.9	**(n=341) 31.1**	**(n=220) 43.2**

experienced intense disasters (damage of house). Hence, more wealthy and educated people are also more resilient to climate-related disasters than others. Moreover, households located in the vicinity of waterbodies are more likely to experience intense impacts (damage on houses) from flood-related disasters compared to those living further away from such hazards.

To conclude, the adaptive capacity (to take action) of communities in the two assessed communities of Chennai following climate-related disasters is limited to allow them to increase their resilience themselves in relation to the three assessed dimensions (physical, social and economic) to better cope with future climate-related disasters. As a result, the previously affected (damage on house) households are likely to experience again future flood-related disasters as their resilience has not improved following past experiences.

This conclusion points out that context-specific solutions are required to enhance the resilience of those parts of a community who have lower coping capacities to anticipate, manage and recover (adaptive capacity) from climate-related disasters.

LOCAL STAKEHOLDER WORKSHOP

The findings from the household surveys in the two wards of Chennai make clear that communities are highly complex entities. Even within a small area (around 1 km^2), disasters (floods) are experienced differently, depending on the location of major natural hazards.

These complex characters of communities also highlight that without the integration of local knowledge urban planning for disaster resilience cannot be achieved. This opinion is supported by other scholars such as Few (2003) and Jabeen, Johnson, and Allen (2010) who encourage the involvement of communities into local decision-making processes. Thus, a local stakeholder workshop was organised as a follow-up event to gain perceptions from local residents and governmental officials.

Local Stakeholders Workshop: Overview

After the household survey was analysed, a local stakeholder workshop took place in Ward 131 during August 2011. The key aim of this workshop was to gather local (perceptions from residents and local government officials) and expert knowledge (academia) to understand and formulate recommendations (actions) on how the resilience of communities can be enhanced to climate-related disasters. This workshop was undertaken in mutual collaboration between the Corporation of Chennai, University of Madras (UoM) and Kyoto University. The selected venue was a school (see Fig. 6) at MGR Nagar (Ward 131), Jafferkhanpet, in Chennai.

While the aim of the workshop was to gain local knowledge on how enhance the resilience at the local level, the following objectives defined the expectations from this event:

- To discuss with representatives from the local public, city officials (from various departments) and local NGOs how the climate-related disaster resilience of the Ward 131 can be enhanced.
- To formulate pathways to a selected number of actions from the AoRA based on the perceptions of local residents.
- To provide a platform for local residents and governmental officials to exchange views on how communities, such as Ward 131, can be liberated from climate-related disaster risk.

Participants of Workshop

As the aim of the workshop was to bring together local residents, local government officials and academicians, the following number of people participated in this one-day workshop: 34 local residents (Fig. 7), 9 local government officials and 19 academicians (mostly students from UoM and Kyoto University).

Fig. 6. Location of Stakeholder Workshop at Ward 131 in Chennai.

Fig. 7. Local Residents at Stakeholder Workshop.

Furthermore, the Commissioner of the Corporation of Chennai and Vice-Chancellor of the UoM inaugurated this workshop (Fig. 8).

Workshop Activities

Based on the CDRI framework and formulation of action measures in the AoRA, the content of this workshop followed this structure. However, since a CAP (presented in Chapter 8) was already developed for the entire city of Chennai, overlapping actions from the AoRA and CAP were identified to be discussed (see Fig. 9) with local residents.

Hence, local residents were requested to accomplish the following tasks through internal discussions:

1. To decide on a total of 31 actions from all the five CDRI dimensions whether they should or can be implemented in the short (up to 2 years), medium (2–5 years) and long-term (more than 5 years).

Fig. 8. Commissioner (Left) and Vice-Chancellor (Right) at Stakeholder Workshop.

Fig. 9. Focus Group Discussions at Stakeholder Workshop.

2. How these actions (mechanisms) can be actually implemented, including who are the stakeholders that are required in the implementation process and what level of responsibility they should have.

In this process, two groups were formed to discuss on 15, respectively, 16 actions from the total 31 identified actions. Despite the large number

(34) of present local residents in the morning session, the participation in the actual focus group discussion was limited. Thus, one group consisted of five local residents and three students from the UoM and the other group had three local residents and four students from UoM. The represented local residents had mixed power in their community. While some residents belonged to a community-based organisation (e.g. residential welfare association) others were normal residents with no specific affiliation.

Outcome of Workshop Activities

Table 6 shows the identified overlapping actions between the AoRA and CAP. Based on the perceptions of local residents 27 out of 31 actions are thought to be implementable within a 2-year horizon. While issues related to disaster management plan-making or enforcement of laws (building permits) and strategies (slum-removal) are believed to take up to 5 years. In contrast, not shown here, the CAP formulated by the Corporation of Chennai foresees for many actions a longer time-span to become fully implemented. This implies that expectations from local residents deviate clearly from how the local government beliefs actions related to DRR can be implemented. Therefore, the results in Table 6 also highlight to some extent the expectations local residents have on implementing DRR as soon as possible. These issues are further discussed in Chapter 8.

Table 7 shows how the local residents perceived different actions should be implemented. Similar to the results in the AoRA a coordinated approach involving different stakeholders (multi-stakeholders) are highlighted in the mechanisms of implementation of many actions. Local residents also perceive that in 29 out of the 31 proposed actions increased attention is required to enforce and improve existing measures. Thus, a stronger presence of the local government to control nuisances in different sectors is sought to be crucial. For example, the enforcement of building codes, but also the actual implementation of a slum-removal strategy as well as strict control of environmental policies.

Apart from tighter enforcement of laws and policies, the local residents expressed their desire for increase cooperation and coordination between themselves and the local government to carry out a number of actions, such as raising awareness about solutions (e.g. solar panels, eco bulb) that would reduce the use of electricity, trainings of emergency teams, solving the absence of proper drainage systems within roads and creating voluntary groups for relief operations following a disaster. Those measures are sought

Table 6. Prioritisation of Overlapping Actions from AoRA and CAP, Done by Local Residents.

Dimensions	Parameters	No	Actions	S (<2 years)	M (2–5 years)	L (>5 years)
Physical	E	1	Promotion of alternative energy sources for owners of buildings/houses (e.g. solar panels)	✓		
		2	Promotion of low energy appliances (e.g. *eco bulb*)	✓		
	W	3	Establishment of *emergency backup* of safe water supply for all citizens for a minimum period of 3 days to provide supply after a disaster	✓		
	S	4	Conduction of *sanitation education* at the community level	✓		
		5	Development of *designated sites* for *debris waste collection* after a disaster	✓		
		6	Introduction of *waste segregation practices* at household/community level	✓		
	R	7	Develop all roads with adequate *drainage systems* to avoid water logging after intense rainfall events	✓		
	HL	8	Strict enforcement of *building permits* for all types of hazards for new buildings		✓	
		9	Reducing the *number of settlements living on hazardous grounds* (relocation, provision of alternative housing)		✓	
Social	P	10	Population control measures to optimise the *population density* at ward level		✓	
	H	11	Initiation of campaigns to raise awareness about *health concerns* (e.g. waterborne diseases) at the household/community level	✓		
	E&A	12	Conduction of *disaster drills* at ward level at least once a year	✓		
		13	Campaigns/awareness programmes *sensitising* the people about the threats of climate change	✓		
	SC	14	Stronger *involvement* of communities in *decision-making processes* (e.g. major projects, development plans, etc.) at ward-level		✓	
	CP	15	Creation of *evacuation plans* at the community-level	✓		
		16	Establishment of *community-led voluntary groups* at ward level for relief and recovery works (Community Emergency Response Teams)	✓		
Economic	E	17	Initiation of skill trainings for urban poor people	✓		
	E	18	Development of networks among different partners to establish a *job platform*	✓		
	B&S	19	Increase of *budget* targeting *disaster risk management* to 3% of overall Corporation budget by 2015	✓		

Institutional	M	20	Develop *disaster management plans* which address *all types of hazards* (multi-hazard risk assessments)		✓
		21	Stronger *involvement of communities in plan-making processes*	✓	
	CM	22	Increase of number of *training courses for emergency teams* to manage future disasters	✓	
	K&D	23	Incorporation of *disaster education* in school syllabus	✓	
		24	Development of localised easy understanding *disaster awareness* materials (*pamphlets*, *videos*, etc.)	✓	
		25	Development of *networks* between schools, communities and other partners to transfer knowledge about the threat of disasters and climate change	✓	
	G	26	Establishment of efficient *multi-hazard early warning systems* using radio, television and SMS technologies)	✓	
Natural	ES	27	Reducing the *discharge of untreated waste* into waterbodies (canals, rivers)	✓	
		28	Increasing area of *urban greenspace* (trees, parks, etc.)	✓	
		29	Campaigns/awareness programmes for communities to inform about the consequences of burning fossil fuels → *reducing air pollution*	✓	
	LU	30	Establishment of sustainable urban development strategies for a balanced, mixed-use of available urban areas	✓	
	EP	31	Strict enforcement of *Environmental Impact Assessments* for all major development projects	✓	

Table 7. Prioritisation of Overlapping Actions from AoRA and CAP, Done by Local Residents.

No.	Duration (S, M, L)	Actions	Mechanisms of Implementation – How to Implement
1	S	Tamil Nadu Government recently announced to promote solar systems in every house	1. Area Associations should cooperate in bringing the solar systems onto posters, pamphlets etc. 2. Flash news: promoting solar systems through loud speakers in streets.
2	S	A given to government distribute on the public people it low cost or free per ration card use	1. Should be easily available to every nation (govt.). 2. A make it advertisement on the television way to reach at public.
3	S	Corporation should coordinate with Metrowater department	1. Storage of water should be monitored at the time of disaster. 2. Checklist must be made to see that all people get water without any influence (bribe).
4	S	A given fund allotted in make it first and managed in Panchayat board	1. A message passed from government servant to public people. 2. Conduct the school level classes and awareness programmes.
5	S	With the help of Corporation debris has to be removed	Government has to initiate the finding of a proper site and as to collect the waste immediately after a disaster.
6	S	Waste segregation practice should be implemented at the final disposal yard – Perungudi and Kodangaiyur	1. Converting the garbage into briquettes which can be used as fuel. 2. Recycling of non-degradable waste.
7	S	Persisting problem – to be solved immediately	1. Coordination between departments is required. Corporation, CMWSSB, Highways. 2. Suggestions from the local people or Residents Welfare Associations to be taken.
8	M	Building permits – should be made very strict. Demolition – unauthorised and deviated buildings	1. Strict rule must be followed while sanctioning. 2. Strict punishment (fine) must be imposed on those who violate the sanction.
9	M	Government has to give orders to improve those peoples and give them information regarding the dangers	1. Those peoples have to be relocated to safer zone and according to their education qualification jobs has to be assigned. 2. Frequent inspection has to be done. 3. If any of them return, they have to be punished with fine.
10	M	Family planning – one family one child plans amendment in strictly	1. The awareness programmes on the college and university levels. 2. The government employees did not get it more than one child to make it law.

11	S	Required immediately	1. Support from NGOs. 2. Advertisements-water participation system (plant) can be provided at least in streets or important places.
12	S	Very critical – a group should be formed from the community	1. Groups must have all age groups represented. 2. Regular meetings and training should be given – must be called at the time of disaster.
13	S	People need to cooperate in reducing the use of electrical appliances	1. Since climate change occur due to global warming, people should use the current energy supply in a minimal level. 2. Need to plant trees more. 3. With the help of Area Association, information have to be given to local people.
14	M	Information from the government regarding major projects need to be announced through the media	1. Ward Councillors has to conduct a meeting with area association. President take a points and a suggestion implement and plans.
15	S	As early as possible	1. Emergency exits in flats, schools & public buildings. 2. Regular drills, proper campaign.
16	S	Proper training for the volunteers	1. Proper communication and regular meetings.
17	S	Through notice to every house	1. It can be implemented through street dramas.
18	S	It is very critical	1. Government should take care – should give priority – arrange loans. 2. Motivate the Corporation to offer jobs or to support them by giving loans (micro).
19	S	It is essential. There are so many NGOs to help the Corporation of Chennai	
20	M	Already disaster management plans are in use, but people don't know the information	1. Government has to do the needful to give the information to the peoples through media. 2. Disaster management plan – schools should take classes selected to this & it should be in syllabus.
21	S	To be knowledgeable person should be taken only the Planning Commission	1. The government should be making it planning basis from their areas. Particular persons knowledge planning designed in the government.
22	S	Very much needed	1. Periodical meeting & training for various disasters. Disaster management should be made compulsory.

Table 7. *(Continued)*

No.	Duration (S, M, L)	Actions	Mechanisms of Implementation – How to Implement
23	S	Already is there	1. Can give practical training (for disaster relief). 2. Can have mock seminar.
24	S	Government and NGO has to initiate the process	1. Through separate session to schools children & parents. 2. Through seminar, conference and association.
25	S	Conference, seminars, rallies	The government awareness programmes in the college and universities and schools student basis awareness make it.
26	S	Already in place	1. Public announcements by (PAS) – using auto. 2. Alerting neighbours.
27	S	Must be implemented immediately – or else the city will become a toxic zone	1. Strict punishments with fine should be imposed. 2. Must be educated (awareness).
28	S	Shift the IT companies to different major cities	1. People need to stay in their place & work. 2. Family planning should be implemented. 3. Agricultural items price should be increased so that people from village won't migrate to cities and then there will be places for trees and plants.
29	S	Controlled auto, bus, transport and used for purified petrols only	They making oil refineries produced the maximum pure oil and less give in to the transporters. The metro trains will be reduced in their air pollution.
30	S	Should take immediate action	1. Coordination between the Government departments public sectors is necessary. 2. Coordination – committee should be formed by the government before any development scheme is initiated.
31	S	It is necessary – to check all development projects	1. Should have coordination between the departments (e.g. pollution central, Corporation, highways, waterways, public works departments, etc.

to be coordinated between area associations (residential welfare associations, ladies groups, etc.) and the local government to become effective.

In general, local residents advocated for enhance cooperation between the local government and communities (community-based groups) in all dimensions whether about issues related to managing a disaster (relief) to preventive actions that would alleviate or lead to avoiding further disasters.

The local residents further raised the need for increased cooperation between different departments among the local government as well as state government to make effective improvements. For example, when it comes to developing strategies that go beyond one jurisdiction (ward, city, etc.), the different governmental departments are expected to coordinate more strongly together.

While local residents have high expectations on increased cooperation with the local government and improved inner-governmental decision-making, little self-ownership is obtained by them to take the lead in one or the other action. This actually, again confirms the findings from the AoRA where communities are not really regarded as primary leaders to implement certain DRR-related actions, but rather as partners in the implementation process. Despite this restricted attitude of communities to initiate or take primary responsibility for enhancing their resilience, there is some notion that actions related to sensitising the people, recycling of solid waste or ward-level planning should be undertaken by local residents themselves.

Other important stakeholders include schools and NGOs as partners for providing crucial information to pupils about climate change and disasters. Enhancing disaster education at schools is regarded as a very important measure for enhancing the knowledge and awareness of local residents.

Finally, fairer distribution of basic services (e.g. water) to local residents is urged to be often mismanaged and under the influence of corruption (bribing); thus, local residents are calling for enhanced transparency and accountability of the local government and semi-private organisations. The discussion on how to enhance the resilience of communities in Chennai to climate-related disasters is further discussed in Chapter 8.

REFERENCES

Chennai Metropolitan Development Authority (CMDA). (2008). *Second master plan for Chennai metropolitan area, 2026*. Chennai: CMDA.

Cutter, S. L., Barnes, L., Berry, M., Burton, C., Evans, E., Tate, E., & Webb, J. (2008). A place based model for understanding community resilience to natural disasters. *Global Environmental Change*, *18*, 598–606.

Drescher, A., Glaser, R., Pfeiffer, C., Vencatesan, J., Schliermann-Kraus, E., Glaser, S., ... Dostal, P. (2007). *Risk assessment of extreme precipitation in the coastal areas of Chennai as an element of catastrophe prevention*. 8. Forum DKKV/CEDIM: Disaster Reduction in Climate Change 15–16 October, Karlsruhe.

Few, R. (2003). Flooding, vulnerability and coping strategies: Local responses to a global threat. *Progress in Development Studies*, *3*(1), 43–58.

Gallopin, G. C. (2006). Linkages between vulnerability, resilience, and adaptive capacity. *Global Environmental Change*, *16*, 293–303.

Jabeen, H., Johnson, C., & Allen, A. (2010). Built-in resilience: Learning from grassroots coping strategies for climate variability. *Environment and Urbanization*, *22*(2), 415–431.

Revi, A. (2008). Climate change risk: An adaptation and mitigation agenda for Indian cities. *Environment and Urbanization*, *20*(1), 207–229.

Twigg, J. (2007). *Characteristics of a disaster-resilient community: A guidance note*. Disaster Risk Reduction Interagency Coordination Group. Benfield: DFID.

CHAPTER 8

MAKING CHENNAI RESILIENT TO CLIMATE-RELATED DISASTERS

INTRODUCTION

In this chapter, the key findings from Chapters 5–7 are reflected to discuss the challenges in the process of making Chennai resilient to climate-related disasters. The structure of this chapter is as follows: firstly, the key findings from the Climate Disaster Resilience Index (CDRI), Action-oriented Resilience Assessment (AoRA) and Climate-related Disaster Community Resilience Framework (CDCRF) are revealed; secondly, a Climate Action Plan (CAP) is presented which aims to serve as a plan to mainstream Disaster Risk Reduction (DRR) in Chennai; thirdly, the initiation of the Safer Chennai Campaign as part of the UN World DRR Campaign is presented; fourth, the implications of the current urban expansion of the Corporation of Chennai is discussed; and finally, further suggestions are made on how Chennai can become resilient to future climate-related disasters.

KEY FINDINGS FROM THE RESILIENCE ASSESSMENTS

In the process of making Chennai resilient to climate-related disasters, the applied CDRI, AoRA and CDCRF served to get an understanding on the functioning of urban communities in the context of Chennai. Thus, Fig. 1 illustrates the sequential approach that begins with a primary assessment to understand resilience at a broader scale before going into more detail to understand how communities at the ward level are functioning.

Followed by such assessments a CAP may serve as a viable action- and implementation-oriented tool for making a city, in this case Chennai, resilient to climate-related disasters. Therefore, in the following sections the key findings for each of the three assessments are summarised.

Climate Disaster Resilience Index

Key findings from the CDRI assessments are as follows:

- There is a strong correlation between the physical resilience of the 10 zones and past population growth rates. This means the higher the physical resilience of a zone, the higher its population growth rate is. This result implies that urbanisation is a positive factor for the physical resilience of an urban area.
- Aspects of livelihood (social cohesion, economic condition of households, physical capacity [housing, basic services]) are strongly reflected in the CDRI which entails that higher wealth or a better economic condition of communities overall increases their resilience. This finding is further supported by the significant correlation between the social and natural dimensions which says, for example, that higher social capital also leads to enhanced compliance to environmental policies. This finding is crucial to engage communities into the process of implementing a CAP which aims to be mostly based on soft-related adaptation or DRR measures.
- The results in the institutional dimension are more or less equal in the 10 zones which point out that aspects of livelihood and urbanisation have no impact on the quality/resilience on how the local authorities are prepared to face and respond to climate-related disasters.

Overall, these key findings suggest that communities with high social cohesion and capital are available and ready to contribute to process of making Chennai resilient to climate-related disasters.

Action-Oriented Resilience Assessment

Key findings from the AoRA are as follows:

- The local government is the actor which has highest responsibility to implement almost all the proposed DRR-related actions (62 out of 63). This implies that the Councillors (respondents) do not consider devolving the principal responsibility of DRR to other stakeholders.

- The results, however, also highlight that a multi-stakeholder involvement is regarded as the most viable approach for the implementation of any of the 63 actions or more simply DRR. Therefore, depending on the nature of the action private organisations, academia, NGOs or communities are sought to become key partners for the local government in the implementation of actions.
- Councillors prioritise those actions to be implemented first which are related to raising awareness (safe water use, community activities, skill training [job]), enforcement of laws (building codes, EIA) and creating the financial capacities to prevent and manage a disaster (insurance schemes, DRR budget of local government). This means that a combination of actions related to raising knowledge among communities and providing the financial capacities are required to enhance the resilience of Chennai's communities to climate-related disasters.

Climate-Related Disaster Community Resilience Framework

Key findings from the CDCRF are as follows:

- Households who have been affected by disasters (house damaged) are to a large extent located along natural hazards (canals, rivers).
- Households who have been affected by a disaster (flood) in the past do not learn or take action from such events. This fact may be explained due to their lower education level (lower ability to speak English [indication of wealth], lower knowledge about disasters and climate change), access and satisfaction of basic services and economic wealth. In other words, households who have been affected by past disasters are mostly less privileged and do not have the same capacities/resilience than those households who have not been affected by a disaster in the past. As a result, lower livelihood conditions increase the susceptibility of people to be affected by a climate-related disaster.
- Although more households have been affected by a disaster in the past in the rapidly urbanising Ward 131 compared to Ward 79 with a stable population (37 per cent in Ward 131 vs. 24 per cent in Ward 79), urbanisation is not per se a driving factor for more people becoming to flood-related disasters. The percentages of households who have been affected by a disaster and also have migrated in the past show no statistical significance between Ward 131 and Ward 79. Therefore, urbanisation does not necessarily lead to more people being exposed to hazards.

In particular, the last finding is important to be further discussed. Since the household survey was conducted among lower- to upper-middle class households, urban poor have been excluded, the often perceived negative connotation towards the term 'urbanisation' needs to be relativised. That means middle-class migration into urban areas must be perceived to be a positive factor to enhance the resilience of an urban area. This finding is confirmed by the CDRI results which indicate that those zones which are located at the outskirts of the city are having higher physical and economic resilience and also higher recent migration of people.

As a result, the initial findings gained from the CDRI assessment are confirmed by the more detailed ward-level assessment. Accordingly, Ward 79 is likely to be more challenged to cope with a future flood-related disaster compared to Ward 131 which is located in a high resilient zone.

CAP FOR CHENNAI

In order to specifically address the challenges of the city to climate-related disasters, a CAP has been developed in joint collaboration with the Corporation of Chennai in 2010. Although the AoRA and CDCRF have not been directly integrated into the formulation process of the CAP, they are of high importance during the actual implementation of the proposed DRR-related action measures that aim to enhance the resilience of Chennai. Accordingly, Fig. 1 again highlights the different steps that lead towards the CAP to reach the ultimate goal of a more disaster resilient city.

Action Planning Process in Chennai

The structure of the CAP follows the five dimensions (physical, social, economic, institutional and natural) already adopted in the CDRI and AoRA. In addition, it consists of three time periods in which actions should be carried out, as follows: short-term (Table 1) up to 2 years, medium-term (Table 2) between 2 and 5 years and long-term more than 5 years (Table 3).

An important characteristic of the CAP is that it focuses on soft adaptation measures and excludes hard measures. The reason is that the CAP aims to be feasible in its implementation requiring limited financial investments. This should stimulate an enhanced activism among local

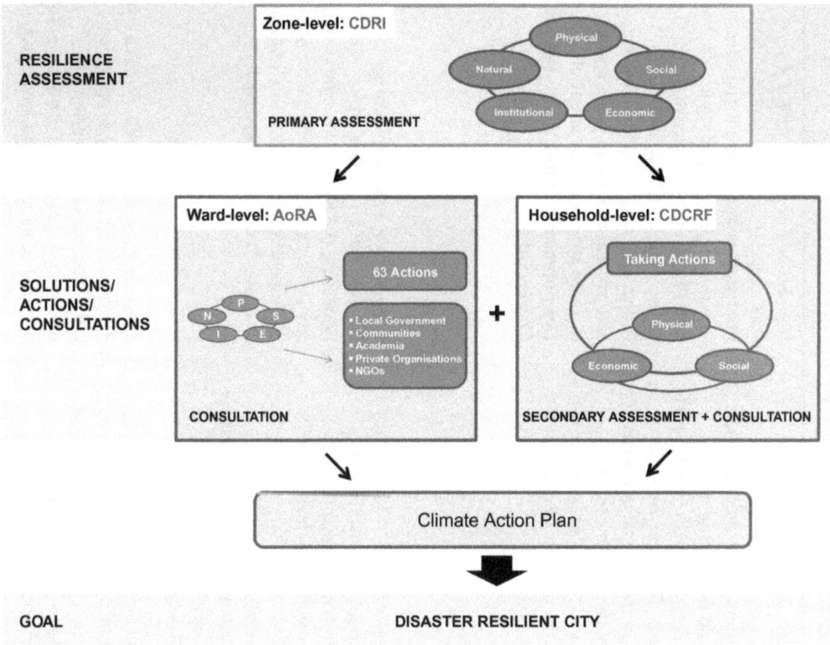

Fig. 1. Conceptual Framework for Climate Disaster Resilience in the Context of Chennai.

stakeholders to take immediate action and thus, become more feasible to become actually implemented.

Thus, Fig. 1 conceptualises the various steps leading to a disaster resilient city. At current, the CAP is still in the process to be formally adopted by the legislative body (Council) of the Corporation of Chennai. It is imminent that the CAP gets legislative formal approval to become effective.

Chennai CAP

In the following three tables the specific actions for the three most important parameters for each dimension and implementation period are highlighted. Additionally, the actions indicate which task from the HFA and Checkpoint from the 'Making Cities Resilient' campaign, described in Chapter 4, they would support.

Table 1. Short-Term CAP Actions: Up to 2 Years Implementation Period.

Physical	1. *Sanitation and Solid Waste Disposal* (*HFA 3 – Task 9*) a) Providing and construction of more public conveniences in slums to prevent open air defecation. Need to educate the people about cleanliness of the place and environment and good sanitation habits. *Benefits Checkpoint 6* b) Encourage people for source segregation of solid wastes. Communities ought to be encouraged to recycle the biodegradable waste by setting up a composite yard wherever possible. *Benefits Checkpoint 1*	2. *Accessibility of Roads* (*HFA 4 – Task 14*) a) Improvements to the existing road network and developing new transport corridors with proper storm water drains. *Benefits Checkpoint 4* b) Provision of intermodal connectivity to improve the efficiency of the transport system and making it more user-friendly. *Benefits Checkpoint 4*	3. *Housing and Land-Use* (*HFA 4 – Task 15*) a) Provision of alternative living areas to relocate people residing in waterways, encroachments, and slums. *Benefits Checkpoint 6* b) Proper planning of urban land-use and provision of more green spaces through stricter enforcement of Development Control Rules. *Benefits Checkpoint 6*
Social	1. *Population* (*HFA 4 – Task 13*) a) Intensify efforts to educate the poor and the most vulnerable people about family planning. Encourage people to have small family by giving incentives. *Benefits Checkpoint 7* b) Educate women about the safe use of family planning tools. *Benefits Checkpoint 7*	2. *Social Capital* (*HFA 3 – Task 10*) a) Conduct mock drills for school children/communities at regular interval by involving them. *Benefits Checkpoint 9* b) Preparation of a localised early warning system tailored to the nature of community/threat they face. *Benefits Checkpoint 9*	3. *Community preparedness during disaster* (*HFA 3 – Task 9*) a) Educating the people through media, handbills, skids, road shows about the impending threat due to climate changes. *Benefits Checkpoint 7*
Economic	1. *Employment* (*HFA 4 – Task 17*) a) Encourage industries to provide need-based skill training programs to the local people and further creation of green employment opportunities. *Benefits Checkpoint 10*	2. *Finance and Savings* (*HFA 4 – Task 17*) a) Extension of bank credits for women self-help groups. Setting up a centre with the purpose of identifying, training and marketing products developed by women self-help groups. *Benefits Checkpoint 10*	3. *Budget and Subsidies* (*HFA 1 – Task 4*) a) Provision of budgetary support for DRR activities and creation of resources for regular contributions to the fund at national and local level. *Benefits Checkpoint 2*

		b) Setting up of regular skill training programs for women self-help groups. *Benefits Checkpoint 10*	
Institutional	1. *Mainstreaming of DRR and CCA* (*HFA 5 – Task 20*) a) Incorporation of rules or modification rules with respect to DRR and stricter enforcement of rules with the help of other stakeholders. *Benefits Checkpoint 1* b) Integration of Disaster Management in school curriculum. *Benefits Checkpoint 7*	2. *Knowledge Dissemination and Management* (*HFA 1 – Task 3*) a) Create awareness through media, skids, handbills and road shows about disasters and also train the communities in first aids and how to act during any disaster. Conduct frequent mock drills and the safety aspects and precautionary acts to be taken up during any disaster. *Benefits Checkpoint 9* b) Train basic health workers in disaster preparedness. *Benefits Checkpoint 9*	3. *Good Governance* (*HFA 4 – Task 19*) a) List of resources available in the area, vulnerability map with indicators with a list of old age people, pregnant women, handicapped and the infants. Understanding the situation for a common approach. *Benefits Checkpoint 5* b) List of voluntary groups with specific task assigned to each of them, to act during an emergency. *Benefits Checkpoint 1*
Natural	1. *Ecosystem Services* (*HFA 4 – Task 12*) a) The impacts due to fast urbanisation and growing vehicle movements lead to severe pollution of the air. Stricter emission controls and encouragement of people to use public transport are primary actions. Also, a good intermodal transport will encourage people to use public transportation. *Benefits Checkpoint 8* b) Prevention of untreated disposal of industrial waste and sewage into water bodies. *Benefits Checkpoint 8*	2. *Land-Use* (*HFA 4 – Tasks 12 & 14*) a) Regulating the land-use, and forming policies with various stakeholders to protect the city's fragile biodiversity. *Benefits Checkpoint 3* b) Protecting the groundwater table through development control rules. *Benefits Checkpoint 3*	3. *Enforcement of Environmental Policies* (*HFA 4 – Task 12 & 14*) a) Enforcement of environment policies and Costal Zone Regulations and creating awareness among people in maintaining the environment. *Benefits Checkpoint 8* b) Creating more open space by developing parks etc. and maintain such open spaces as green space. *Benefits Checkpoint 8*

Table 2. Medium-Term CAP Actions: Between 2 and 5 Years Implementation Period.

Physical	1. *Electricity* (*HFA 4 – Task 17*) a) Promotion of alternative source of energy like solar panels and encourage using energy saving equipments (CFL Lamps, equipments with star rating). *Benefits Checkpoint 1* b) Educate the people to save energy and make it a habit through media and handbills etc. *Benefits Checkpoint 1*	2. *Accessibility of Roads* (*HFA 4 – Task 14 & 7*) a) Provide all roads with adequate storm water drains and interlink the drains to canals to prevent water logging during flood. Clean and de-silt the waterways regularly to prevent water logging during rain. *Benefits Checkpoint 4* b) Preparation of hazard maps indicating the low-lying areas and water logging areas. *Benefits Checkpoint 4*	3. *Housing and Land-Use* (*HFA 4 – Task 15*) a) Revising/modifying the National Building Code to incorporate the recent developments/requirements with the involvement of the Institutions, Communities, NGOs and various stakeholders, and enforcement of the same strictly. Amending the rules for stronger enforcement of environmental policies. *Benefits Checkpoint 3* b) Identifying the areas and demarking the same for future developments for open space or green area like parks. *Benefits Checkpoint 6*
Social	1. *Population* (*HFA 4 – Task 13*) a) Providing the basic amenities like good sanitation facilities, safe drinking water, good interior roads, schools, parks, play fields, community welfare centres and health posts for the urban poor living in slums to change the characteristic of slums in the next 2–5 years' time. *Benefits Checkpoint 6*	2. *Education and Awareness* (*HFA 3 – Task 9*) a) Creating awareness by including DM in school curriculum to inform about threats due to climate changes, waterborne diseases during flood, and safety precautions during the monsoon period. *Benefits Checkpoint 7*	3. *Social Capital and Community Preparedness* (*HFA 1 – Task 2*) a) Active involvement of people/community in decision-making process for a systematic coordination to reduce the risks due to disasters. *Benefits Checkpoint 1* b) Recovery plan after any disaster by having a common solution to address the requirements of the local community. *Benefits Checkpoint 10*

Economic	1. *Income* (*HFA 4 – Task 16*) a) Provide skill training for the urban poor/women to seek alternative opportunity during any disaster to sustain their life like handicrafts, with the involvement of private parties and buy back arrangements or marketing tie-ups. *Benefits Checkpoint 10*	2. *Finance and Savings* (*HFA 4 – Task 17*) a) Creation of funds for emergency assistance and extension of micro credit or repayment holiday after any disasters. *Benefits Checkpoint 10* b) Encourage people to save/insure their produces or properties which will help them during an emergency. *Benefits Checkpoint 10*	3. *Budget and Subsidy* (*HFA 1 – Task 4*) a) Provide separate funds for disaster mitigation and management plans and with sustained efforts create awareness among the society for disaster preparedness. *Benefits Checkpoint 9*
Institutional	1. *Mainstreaming of DRR and CCA* (*HFA 1– Task 2*) a) Develop a disaster management plan and CAP with the help of past experience, and the lessons learnt from other cities/nations to address all types of hazards with the help of local community, NGOs and voluntary organisations. *Benefits Checkpoint 3* b) Interact and conduct regular meetings with various stakeholders to update the knowledge which will improve the resilience of the community. *Benefits Checkpoint 1*	2. *Knowledge Dissemination and Management* (*HFA 2 – Task 8*) a) Conduct mock drills at regular intervals and asses the strength and improve the strength and knowledge and conduct programs to children with the help of institutions or volunteers or NGOs who are familiarised in their respective field. *Benefits Checkpoint 9* b) Prepare a plan for networking various groups during any disaster. *Benefits Checkpoint 1*	3. *Good Governance* (*HFA 5 – Task 19 & 20*) a) With latest applications like GIS and remote sensing prepare a comprehensive plan indicating the high rise buildings, vulnerable areas, risk prone areas etc. with respect to various disasters can be made available through internet. Such information can be updated on regular basis for effective monitoring. *Benefits Checkpoint 7* b) Sharing of knowledge with various institutions, and communities and updating the information through regular training/meetings for various stakeholders or through journals or publications. *Benefits Checkpoint 1*

Table 2. (*Continued*)

Natural	1. *Pollution* (*HFA 4 – Task 12*) a) Focus on long-term strategy to reduce the use of fossil fuels and identify the alternative energy. Encourage institutions and private organisations to spend more on Research and Development on creating alternative source of energy. *Benefits Checkpoint 8* b) Create a separate transport corridor for vehicles to the port carrying minerals and ores. *Benefits Checkpoint 8*	2. *Land-Use* (*HFA 4 – Task 12 & 14*) a) Strict enforcement of environmental impact assessments for all multi-storey building or mega structures. Particularly, the sewage disposal and the effect of groundwater due to such multi-storey building need to be assessed. *Benefits Checkpoint 8* b) Urban land zoning should be strictly enforced. *Benefits Checkpoint 6*	3. *Environmental Policies* (*HFA 4 – Task 12*) a) Long-term policy/plan based on the severity, intensity and frequency of natural hazards and formulates an integral approach with institutions, experience and knowledge of other countries and various stakeholders. *Benefits Checkpoint 8* b) Enact law or notify such changes and enforce the same strictly. *Benefits Checkpoint 8*

Table 3. Long-Term CAP Actions: More than 5 Years Implementation Period.

Physical	1. *Electricity* (*HFA 4 – Task 17*) a) Amendment of building rules/codes for energy conserving. Make energy-audit as mandatory. Encourage buildings to save electricity by achieving LEED standards. *Benefits Checkpoint 4* b) Education of people to save energy and encourage the use of star rating equipments. *Benefits Checkpoint 1*	2. *Water* (*HFA 5 – Task 12 & 20*) a) Provision safe water to everyone. Adequate numbers of new catchment areas to be identified to develop as a source of water supply for the growing and future demand. Construction of new tanks to prevent the loss of rainwater during intense rainfall. *Benefits Checkpoint 1* b) Recycling of waste water for reuse and ensure the same for every large buildings and industry. *Benefits Checkpoint 1*	3. *Sanitation and Solid Waste Disposal* (*HFA 4 – Task 12*) a) Provision of a long-term strategy for reducing the solid waste by recycling and reusing it. Identification of alternative land-fill sites for the disposal of solid waste. A separate tax needs to be collected for specific goods (non-recyclable goods) at the time the goods are purchased for disposal purposes of the products. Similar approaches need to be adopted for other sectors of solid waste management. *Benefits Checkpoint 7* b) Ensure the prevention of outbreak of diseases during flood, rain, etc. and make mandatory the immunisation as a regular one and bring everyone under the immunisation programme. Development of a preventive mechanism to avoid the spread of diseases to other communities. *Benefits Checkpoint 7*
Social	1. *Population* (*HFA 4 – Task 13*) a) Development of new areas and housing constructions with basic amenities for relocating the people and decongestion of the thickly populated areas. Indemnification that	2. *Health* (*HFA 4 – Task 13, 19 & 20*) a) Development of an action plan to eradicate diseases through regular immunisation programs and networking of health posts which serve as key agency in obtaining the	3. *Social Capital and Community Preparedness* (*HFA 1 – Task 1*) a) Institutionalisation of Disaster Management with various stakeholders and update their knowledge. Conduct regular meeting

Table 3. (*Continued*)

	basic amenities are provided. *Benefits Checkpoint 6* b) Creation of industries in new identified areas to provide employment opportunities to sustain their livelihood. *Benefits Checkpoint 10*	necessary information regarding the availability of drugs and medical assistance during an emergency. *Benefits Checkpoint 9* b) Improvement of rescue and relief methodologies with the help of paramedical staff and provision of updates on regular basis. *Benefits Checkpoint 9*	and review the safety aspects. *Benefits Checkpoint 1* b) Demolishment of old (vulnerable) buildings and retrofitting of heritage buildings with the help of authorities to ensure the safety of the community. Long-term plan for changing the huts/thatches/sheds and vulnerable buildings through financial assistance and subsidies. *Benefits Checkpoint 2*
Economic	1. *Employment* (*HFA 4 – Task 16*) a) Encouragement of industries and academia to develop a curriculum for people having without formal education to create employment opportunities to sustain their livelihood. Private industries and the association of industries should be attracted to develop training programs in mutual cooperation with the Chennai Corporation. *Benefits Checkpoint 1* b) More educational institutions with specific courses for people without any formal education need to be developed to improve their skills. *Benefits Checkpoint 1*	2. *Finance and Savings* (*HFA 4 – Task 17*) a) Government needs to encourage people to enhance their tax saving habits by offering tax free long-term bonds. Higher interest rates to encourage savings should be an option. *Benefits Checkpoint 10* b) General contributions to Calamity fund may be encouraged to be used during an emergency. *Benefits Checkpoint 10*	3. *Budget and Subsidy* (*HFA 1 – Task 4*) a) Institutionalisation of DRR in all developmental activities. Separate funds need to be considered to be allotted for initiatives which aim to reduce disaster risks. *Benefits Checkpoint 1* b) For retrofitting of old buildings or removal of huts and thatched roofs, subsidies need to be given for safeguarding the communities. Sufficient funds need to be provided in the budget. *Benefits Checkpoint 4*
Institutional	1. *Mainstreaming of DRR and CCA* (*HFA 5 – Task 19 & 20*) a) Development of a long-term disaster management plan to suit the	2. *Institutional Collaboration* (*HFA 2 – Task 8*) a) Long-term plan to be evolved with the help of various stakeholders and	3. *Good Governance* (*HFA 5 –Task 20*) a) A comprehensive study needs to be conducted regularly about the locality, safety aspects, impact of

	increasing threats with the help of institutions, agencies involved in such activities, and through the knowledge acquired from other people/countries/regions. Implementation of a mechanism for regular monitoring and reviewing of the Disaster Management Plan with the inputs from various stakeholders, like communities. *Benefits Checkpoint 1* b) The threats due to various hazards and the impact on civil society should be communicated to the community regularly through sustained efforts. *Benefits Checkpoint 9*	the same may be strictly enforced with the regular feedback from the communities and the stakeholders. *Benefits Checkpoint 1*	environment due to various developmental activities, population, vulnerability with respect to their past experience, available infrastructures, etc., and the strengths and weaknesses of the communities in particular needs to be analysed. With such data a more localised disaster management plan may be arrived. *Benefits Checkpoint 4* b) Reduction of risks faced by communities in developing a specific long-term strategy for the communities highlighting preventive solutions for a period of time. *Benefits Checkpoint 4*
Natural	1. *Intensity and Severity of Hazards* (*HFA 2 – Task 5*) a) The frequency and severity of climate-related hazards need to be studied to develop a comprehensive plan (including hazard maps) for DRR where the impacts of each hazards are put in relation to the urban risks. *Benefits Checkpoint 1* b) A comprehensive early warning system needs to be evolved for community preparation. *Benefits Checkpoint 9*	2. *Ecosystem Services* (*HFA 4 – Task 12*) a) Provision of a long-term strategy to achieve Zero pollution level by reducing the emission level to zero and treating the waste from households and industries to 100%. *Benefits Checkpoint 8* b) Cleaning of waterways and interlinking of waterways to achieve pollution free water bodies and increasing the water carrying capacity during flood events. *Benefits Checkpoint 4*	3. *Environmental Policies* (*HFA 4 – Task 12*) a) Relocation of high polluting industries, prevention of multi-storey buildings in developed areas to reduce population density and the impact on the environment. *Benefits Checkpoint 8* b) Creation of more green spaces in the new developing areas and planting of more trees along the waterways declaration and designation of protected areas. *Benefits Checkpoint 8*

Identified Stakeholders for CAP Implementation

Following the formulation of the CAP, the specific key stakeholders are identified that would be involved in the implementation of actions.

a) *State Government Departments*: Municipal Administration and Water Supply Department, Revenue Department, Home Department, Finance Department, State Planning commission, Ministry of Housing and Urban Development, Ministry of Highways, School Education Department and Ministry of Public Works Department;
b) *State Government Agencies*: Chennai Corporation, Chennai Metropolitan Development Authority, Chennai Metro Water Supply and Sewerage Board, Directorate of Municipal Administration, Tamil Nadu Slum Clearance Board, Metropolitan Transport Services, Directorate of Public Health Administration, Directorate of School Education, Directorate of Town and Country Planning, Directorate of Tamil Nadu Fire and Rescue Service, Chennai Collectorate, Tamil Nadu Pollution Control Board, Commissioner of Revenue Administration, Commissioner of Police, Greater Chennai and Commissioner of Police, Chennai Suburban Police;
c) *State Government Departments*: Municipal Administration and Water Supply Department, Revenue Department, Home Department, Finance Department, State Planning Commission, Ministry of Housing and Urban Development, Ministry of Highways, School Education Department and Ministry of Public Works Department;
d) *Central Government Departments*: Ministry of Home Affairs, Ministry of Urban Development and Ministry of Finance;
e) *Private Sector*: Confederation of Indian Industries and Association of Small and Medium Scale Industries;
f) *Non-Government Organisations*: Exnora International, and Various NGOs;
g) *Academia*: Regional Metrological Department, University of Madras and Anna University, Public and Private Schools;
h) *Community-Based Associations*: Residential Welfare Associations and Women Associations.

Way Forward for Fund Mobilisation

The Corporation of Chennai can act as a nodal agency to formulate a draft policy for the CAP. With the input from various stakeholders it can develop a draft plan and invite suggestions from the public by conducting public hearings and before submitting it to the Council for approval. Further this can be forwarded to Government for approval. Once this is approved, necessary budget allocations may be provided by the Government. To create a Fund for execution of the CAP, in each and every developmental project 0.5–1 per cent are proposed to be set aside towards this. This may be appropriated as a separate fund which can be utilised for implementing the CAP in a systematic manner. Also, a separate wing needs to be formed within each department with specific tasks and roles in line to the Disaster Management Act. With specific task and role, the achievements can be monitored and reviewed to assess the impact of the role played in each department in relation to creating awareness. Additionally, a separate committee needs to be established to review the progress. Every project submitted for approval it should have certain parameters to support the policies of the CAP.

The need of the hour is to create awareness and educate the communitics to gain support for the implementation of the CAP.

Funding Pattern of Corporation of Chennai during Non-Disaster Times
The matrix shown in Fig. 2 explains graphically how the Corporation of Chennai acquires funds for development projects. The implementation of the CAP may follow this structure and make use of all the four, respectively five, key sources for financing proposed actions.

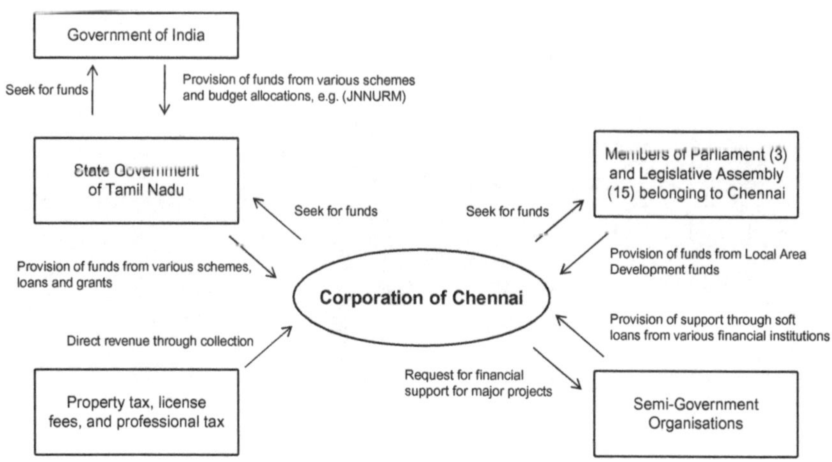

Fig. 2. Fund Mobilization for Corporation of Chennai during Non-Disaster Times.

SAFER CHENNAI CAMPAIGN

On 19 August 2010, the Corporation of Chennai launched (see Fig. 3) the Safer Chennai Campaign in a function held at the Chepauk Campus at the University of Madras to officially demonstrate the city's participation in the United Nations International Strategy for Disaster Reduction (UNISDR) 2010–2011 World Disaster Reduction Campaign: Making Cities Resilient (see Chapter 4). During this event the Mayor and Commissioner of the Corporation participated as well as officers from the State Government and academicians from University of Madras and Kyoto University.

In the formal application process for this campaign, the Corporation of Chennai evaluated itself based on the 10 criteria about its current status, which is shown in the table above. Looking again at the actions formulated in the detailed CAP, indications are given which checkpoint of the list above will benefit if a particular action gets fully implemented.

As of 2010, a major challenge for Chennai is to create an adequate funding scheme that would secure a recurrent yearly budget for DRR. Although the city is confident to have well-functioning early-warning systems (checkpoint 9) thanks to the recently introduced sms-warning system and occasionally conducted drills, the city is not yet fully prepared to provide sufficient emergency management, according to the various assessments presented in Chapters 5–7.

In particular, there is not yet a comprehensive disaster management plan created that would support other than local governmental staff during an emergency. Thus, the city has a contingency plan which is, however, used largely for internal coordination.

Fig. 3. Launching Event of Safer Chennai Campaign, 19 August 2010.

To conclude, the self-assessment by the Corporation of Chennai underlines the need for enhanced actions to which the CAP would be a primary tool to set a guideline what kind of actions should be undertaken during which length of period.

Ten-Point Checklist Linked to CAP

As Tables 1–3 show the entire CAP, each proposed action is attributed with an indication to which checkpoint (Table 4) it is expected to provide

Table 4. Chennai's Ten-Point Checklist.

Ten-Point Checklist – Essentials for Making Cities Resilient[a]	Status[b]
1. Put in place *organization and coordination* to understand and reduce disaster risk within the local government, based on participation of citizen groups and civil society-build local alliances. Ensure that all departments understand their role and contribution to disaster risk reduction and preparedness.	2
2. *Assign a budget* for disaster risk reduction and provide incentives for homeowners, low-income families, communities, businesses and public sector to invest in reducing the risks they face.	1
3. Maintain up-to-date data on hazards and vulnerabilities, *prepare risk assessments* and use these as the basis for urban development plans and decisions. Ensure this information and the plans for your city's resilience are readily available to the public and fully discussed with them.	2
4. Invest in and maintain *critical infrastructure that reduces risk*, such as flood drainage, adjusted where needed to cope with climate change.	2
5. Assess the *safety of all schools and health facilities* and upgrade as necessary.	2
6. Apply and enforce *realistic, risk-compliant building regulations and land-use planning principles*. Identify safe land for low-income citizens and develop upgrading of informal settlements, wherever feasible.	2
7. Ensure *education programmes and training* on disaster risk reduction are in place in schools and local communities.	2
8. *Protect ecosystems and natural buffers* to mitigate floods, storm surges and other hazards to which your city may be vulnerable. Adapt to climate change by building on good risk reduction practices.	2
9. Install *early warning systems and emergency management* capacities in your city and hold regular public preparedness drills in which everyone participates.	3
10. After any disaster, ensure that the *needs of the survivors are placed at the centre of reconstruction* with support for them and their community organizations to design and help implement responses, including rebuilding homes and livelihoods.	2

[a]The checklist builds on the priorities identified in the Hyogo Framework for Action 2005–2015: Building the Resilience of Nations and Communities to Disasters – www.unisdr.org/hfa.
[b]Status levels: 1, poor/nothing in place; 2, some progress in place; 3, in place, well-functioning or N/A.

benefit. Thus, it can be summarised that all checkpoints have been addressed which indicates that the CAP can be perceived to be in line with international aspirations on making cities resilient to disasters. In particular, actions related to building organisation and coordination for effective DRR implementation have been defined as well as solutions related to the protection of the natural environment. The issue of assigning a budget (checkpoint) for DRR is a difficult task, but has been integrated in the short- and long-term measures. As the JNNURM has been established

URBAN EXPANSION OF CHENNAI

Largely due to Chennai's urban growth, the Corporation of Chennai has been expanded in September 2011 from 176 to 426 km^2 (The Hindu, 2011). Also new zones and wards (new boundaries) have been created based on updated population levels (see Fig. 4). Thus, each ward should have a more or less equal population. This is important as the now 200 wards (within 15 zones) are each of them represented by one Councillor. Accordingly, every ward is supposed to have around 45,000 people and 35,000 voters, according to The Hindu (2011). As a result, the former Corporation of Chennai received 107 wards which are more or less matching the city's current population. Accordingly, the new Chennai is close to double its previous population size. The reasons and benefits for this urban expansion, which follow other example in India, such as Greater Mumbai, Greater Bangalore, Greater Hyderabad and New Delhi, is to trigger more resources, optimise expenditure and improve the administrative services (The Hindu, 2011). Although the new expanded Chennai now includes major parts of the Greater Metropolitan Area of Chennai, Chennai's urban area is likely to continue the sprawling into ecologically sensitive and hazard-prone areas.

Fig. 4 points out various areas particularly in the south which are low lying. In these areas new urban growth will take place and require balanced approaches to meet economic development interests versus ecological and biodiversity-related concerns. Failure of commitment to balanced use of land may undermine the various benefits gained from this expansion. In order to achieve balanced land-use, the designation of certain areas where no development is allowed to take place is required particularly, in the low-lying areas in the south of Chennai. These areas are highly susceptible to flooding and since they also include fragile ecosystems (lakes, ponds,

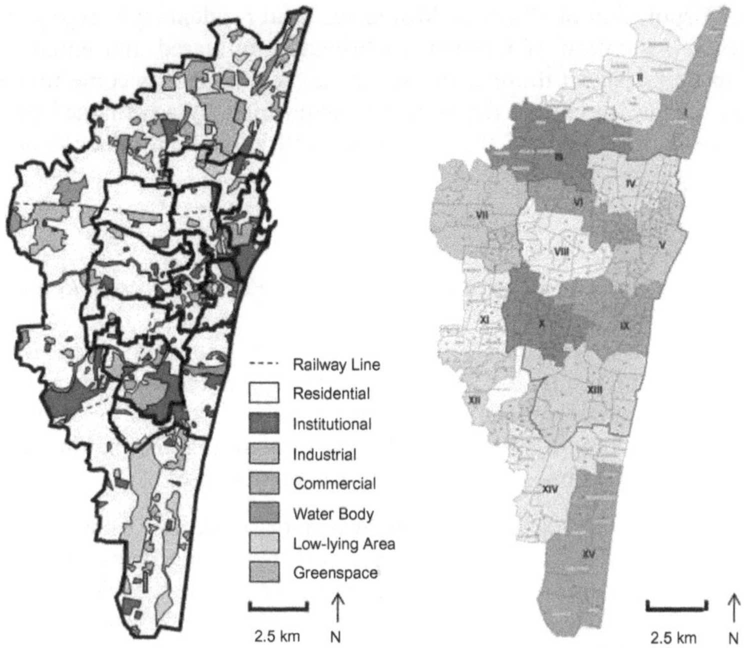

Fig. 4. Land-Use (2006) in Chennai versus Urban Expansion (New Zoning). *Source*: Modified from CMDA (2008) and Corporation of Chennai (2011).

swamps, etc.) measures are required to ensure adequate protection of the environment. Thus, enhanced consideration of environmental aspects into planning decisions is indispensable for delivering sustainable development in the new expanded areas. Furthermore, views from local residents need to be considered in consultation processes of large-scale development plans.

Larger administrative areas ease the planning of infrastructure projects and thus, the expansion has the potential to provide improved planning for providing better transport solutions and basics services (electricity, water, sewerage, etc.) to local residents. Furthermore, this urban expansion has the potential – at least for the next few years – to generate better urban planning of yet non-developed areas through the consultation of local people and actors, mentioned above. On the other hand, a larger administrative body will be required to serve for twice as many people compared to before the expansion, which also increases the level of power at the centre of the Corporation of Chennai. Moreover, local residents will expect to see the new Corporation of Chennai to provide non-biased and equal treatment to every citizen more than ever since 'new land' has come under its control that is susceptible for improper acquisition by advantaged people. Hence the enforcement of the rule of law will likely be a major debate in the coming years.

Linkage to Conceptual Framework for Climate-Related Disaster Resilience

As mentioned before, the expansion of the municipal area of the Corporation of Chennai has direct implications over the jurisdiction of you newly administered land which was previously controlled by numerous smaller municipalities, towns and villages. Linking this fact to the various tools applied in this book, the following key points arise:

- The new expanded area also brought a re-organisation of the ten zones which are now increased to 15 zones with new boundaries. Hence, the CDRI study is unfortunately not anymore directly applicable to the new context. Nonetheless, the findings of this study provide crucial information on the potentials and challenges observed in areas located in the urban fringe. Similar to the re-organisation of the zones, the previous 155 wards have been provided with new boundaries which are in average making each ward bigger as now 200 wards cover an area which is more than twice as big compared to the size before the expansion.
- Despite the fact that many results from the CDRI, AoRA and CDCRF study are not anymore reflecting the new context, they have great potential to be replicated in the expanded Corporation of Chennai. The stepwise approach, formulated in Fig. 1 remains valid.
- The need to generate a database on issues related to resilience is greater than ever in the expanded Chennai as many new potential development areas are of a highly ecological value. Thus, processes of consultation and mutual decision-making with the public and other stakeholders are indispensable.
- Finally, the local government is now handling a much bigger area and higher number of residents which cause further pressures to provide services with limited capacities. This sector requires immediate support through training of staff and transition to perform services by using computer technologies that are replacing the traditional file and folder (hard-copies) system.

FURTHER SUGGESTIONS FOR ENHANCING THE CLIMATE-RESILIENCE OF CHENNAI

Based on the findings from the CDRI, AoRA, CDCRF, various workshops, CAP and the recent expansion of Chennai, Fig. 5 aims to deliver a

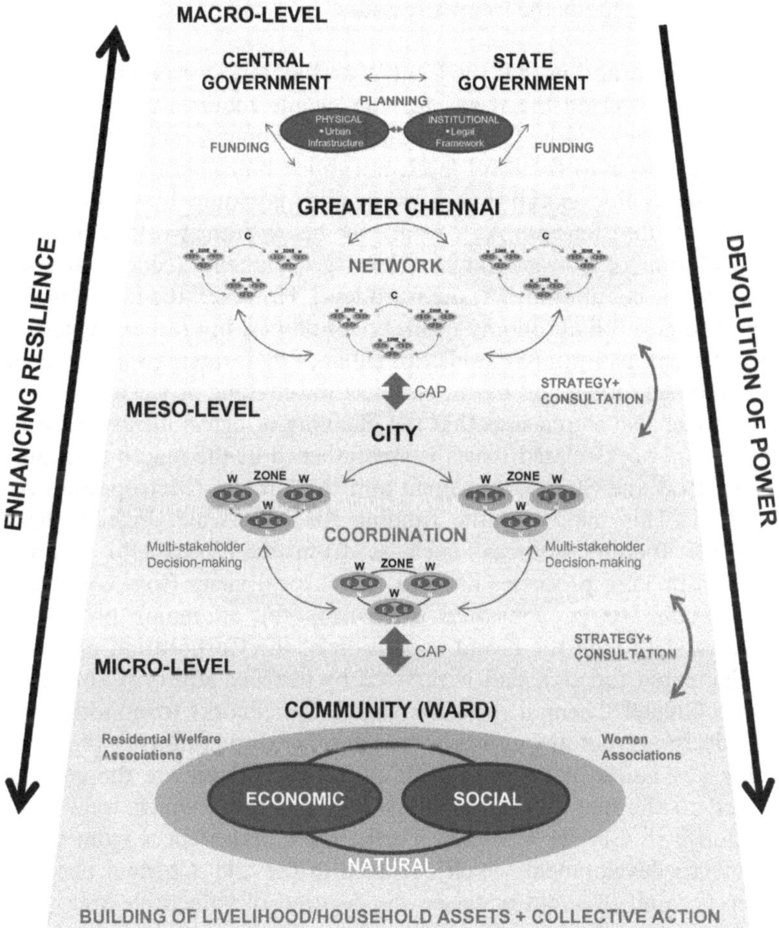

Fig. 5. Framework for Enhancing Climate-Related Disaster Resilience in Urban Communities in Chennai.

framework which could serve for Chennai to enhance its climate-related disaster resilience.

The previous sections have highlighted the challenges which are faced differently between various communities in Chennai, for example, resilience levels are higher in western and southern parts of Chennai and lower in older and less economically vibrant areas. Therefore, the question continues to arise which is how can Chennai's communities become better prepared and enhance their skills to respond to future disasters. In this regard, Fig. 5 illustrates and suggests the following issues:

- The persistent notion that the Central and State Government are the key actors in delivering the needs for the people requires re-thinking. Not only since the 74th Constitutional Amendment, adopted in 1992, devolved power to the local level, but much more because disasters are felt at the local level and thus, local people may know best how to alleviate climate disaster risk. As a result, the below framework advocates for a devolution of power from top level governments (central and state) down to the communities at the ward level. However, the model does not expect to give full autonomy to the communities, but rather suggests that the different institutional levels can enhance their interaction and coordination to smoothen and harmonise decision-making processes.
- The model also emphasises that the planning of urban infrastructure and legal framework-related issues is strengthened at the macro-level among the Central and State government and the Greater (Metropolitan Area) Chennai. This may ease the funding for large-scale projects through available funding sources, such as from the JNNRUM. However, decision-making processes have to reflect local views from communities who require to be consulted extensively for all major projects and changes. Although the recent expansion of the Corporation of Chennai has increased the area that is serviced by just one authority, other cities within Greater Chennai should engage into networks to smoothen planning decisions, but also to support each other during disaster times.
- Other key issues to be dealt with at the micro-level are the challenges related to the new migration of people into areas which were not yet developed in Greater Chennai. Furthermore, experiences from the past of uneven development which occurred in the 'old' Chennai need to be taken account to avoid or lessen the creation of vulnerable areas (pockets) where local disasters may occur. Accordingly, planning of infrastructure to meet future land-use scenarios also need proper legislative frameworks that allow a balanced urban growth and mixed land-use.

These issues need to be considered in planning and policy decisions at the top-level (macro-level). In this regard, the established JNNURM programme with its associated elements (e.g. BSUP) is a viable funding source to achieve this balanced growth of Greater Chennai.

- Considering that sustainable development is a pre-requisite for achieving high resilience to climate-related disasters, concepts related to low economic development are needed to be integrated into future land-use planning and urban growth scenarios of Chennai. In order to allow such kind of development, action needs to be taken at the macro-level by the various stakeholders which are dealing with funding and legislative issues.
- Between the macro- and meso-level or between Chennai City and Greater Chennai, strategies should be developed through consultation processes at the institutional and community-level to develop CAP-related plans and policies that could form a concrete agenda on how resilience is effectively enhanced. The same pattern is suggested between the micro- and meso-level.
- At the micro-level, socio-economic or livelihood aspects are best addressed by residents themselves. Forming a collective power among different residents in communities cannot be steered from higher levels of power. Furthermore, the availability of residential welfare associations and women associations at the community-level serve as ideal Community-Based Organisations (CBOs) that can trigger the needed collective power to improve the livelihood conditions of people. These CBOs are crucial to serve during disaster times as local people have highest trust in people and organisations they are familiar to. By devolving power to such bodies voices and views from local people could be better heard and transferred into decision-making processes. Regarding the natural aspects, many environmental problems, such as the often occurring solid waste disposal into streets could be an issue taken up by the before mentioned CBOs to collectively address this problem and introduce sustainable solid waste management practices that would primarily require the sensitisation of people. Although this problem cannot be solved alone by communities the support from the city government is crucial and thus, a sound relationship between these two actors is a must. However, local CBOs in Ward 131 speak of a constrained relationship with the local government. Lack of trust, coordination and consultation are the major obstacles between these actors which are hindering solutions to be found to these many problems. A potential solution to improve this situation would be through regular meetings between CBOs and the local government (at the ward- and zone-level), according to local perceptions.

The above descriptions for a potential resilience enhancing framework for urban communities in Chennai aimed to integrate aspirations from international guidance Hyogo Framework for Action (HFA) to enhance participatory solutions, but also reflected the current existing hierarchical and centralistic governmental system in India and Chennai.

REFERENCES

Chennai Metropolitan Development Authority (CMDA). (2008). *Second master plan for Chennai metropolitan area, 2026*. Chennai: CMDA.

Corporation of Chennai. (2011). *Expansion of Chennai zone-wise*. Retrieved from http://www.chennaicorporation.gov.in/images/MAP_ZONE_WARD.jpg. Accessed on November 30.

The Hindu. (2011, September 9). Chennai Corporation set to have 45 more wards. *The Hindu*. Retrieved from http://www.thehindu.com/news/cities/Chennai/article2438559.ece. Accessed on November 30.